Integrating Sustainable Agriculture, Ecology, and Environmental Policy

Integrating Sustainable Agriculture, Ecology, and Environmental Policy

Richard K. Olson
Editor

Food Products Press
An Imprint of
The Haworth Press, Inc.
New York • London • Norwood (Australia)

Published by

Food Products Press, 10 Alice Street, Binghamton, NY 13904-1580 USA.

Food Products Press is an Imprint of The Haworth Press, Inc., 10 Alice Street, Binghamton NY 13904-1580 USA.

Integrating Sustainable Agriculture, Ecology, and Environmental Policy has also been published as *Journal of Sustainable Agriculture,* Volume 2, Number 3 1992.

Library of Congress Cataloging-in-Publication Data

Integrating sustainable agriculture, ecology, and environmental policy / Richard K. Olson, editor.

p. cm.

Includes bibliographical references.

ISBN 1-56022-023-6(H) acid free paper – ISBN 1-56022-024-4(S) acid free paper

1. Sustainable agriculture – United States – Congresses. 2. Sustainable agriculture – Congresses. 3. Agricultural ecology – United States – Congresses. 4. Agricultural ecology – Congresses. 5. Environmental policy – United States – Congresses. 6. Agriculture and state – United States – Congresses. I. Olson, Richard K.

S22.I57 1992

338.1'62 – dc20

92-13752

CIP

Integrating Sustainable Agriculture, Ecology, and Environmental Policy

Integrating Sustainable Agriculture, Ecology, and Environmental Policy

CONTENTS

ALL FOOD PRODUCTS PRESS BOOKS
AND JOURNALS ARE PRINTED
ON CERTIFIED ACID-FREE PAPER

ABOUT THE EDITOR

Richard K. Olson is an on-site contractor at the U.S. Environmental Protection Agency's Environmental Research Laboratory (ERL-C) in Corvallis, Oregon. He is responsible for program development and research on the environmental effects of sustainable agriculture, and on the use of constructed wetlands for wastewater treatment. He previously served as Director of the Western Conifers Research Cooperative of the National Acid Precipitation Assessment Program, managing a research program on the effects of air pollution on western forests. Before coming to ERL-C, his areas of research included forest nutrient cycling, canopy/atmosphere interactions, and vegetation patterns and processes in sagebrush ecosystems.

Integrating Sustainable Agriculture, Ecology, and Environmental Policy

Preface

Can ecology aid in developing sustainable agriculture? A fascinating question that is broached by the prestigious authors of this book. Each contributor to *Integrating Sustainable Agriculture, Ecology, and Environmental Policy* has summarized their views concerning this intriguing issue in a concise, easy-to-read format.

This book is the product of a conference sponsored by the U.S. Environmental Protection Agency and the Institute for Alternative Agriculture, where internationally known ecologists, economists, sociologists, soil scientists, and government policy-makers discussed how ecology should and could address agricultural sustainability. These scientists met in a setting which fostered intensive dialogue among participants. Although the debates per se are not published in this volume, the main participants' opinions are clearly presented. Major debate occurred regarding the level of synthesis necessary to bring ecologists and agriculturalists together. Process level to landscape level discussion points were raised. Choice of integration level clearly depended on each person's definition of sustainability.

This book is a formal summary of the major participants' views on issues considered at the conference. For example, how does federal governmental policy affect the environmental sustainability of agriculture? Dr. Otto Doering in his chapter emphasizes that "changes in existing farm programs, or even elimination of them, would not result in large or absolute changes in farming inputs nor in basic ways farmers farm. Nonfarm policies have altered farm structures more than so-called farm programs." Doering challenges us to develop new kinds of policies that fit the desired endpoint.

A new transdisciplinary approach to agriculture is espoused by Dr. Gary W. Barrett. He stresses that a new integration of present research, and educational and management philosophies is necessary. Barrett envisions a new field of study, "agrolandscape ecology," that will aid in "development of a sustainable and cost effective agriculture for future generations."

Dr. Richard Lowrance argues that ecologists need to agree on an organizational level where synthesis can begin, and he carefully defends the watershed/landscape as that level because so much data already exists as a starting point. After choosing that level, we must then measure the "impact of perturbations on the responses of watersheds relative to objective indicators of sustainability." This will require a new research effort and a team approach much like that advocated by Barrett. A large problem seems to be that "objective indicators" for use in assessment are still lacking and need to be developed.

Ms. Sandra Henderson believes that present educational efforts in ecology and sustainable agriculture are focused at the wrong level. She succinctly points out that the "precollege school system (K-12) is the essential place to begin disseminating information that will allow individuals to understand the implications of different agricultural practices in terms of their own health, the health of their environment, and the maintenance of their food supply."

Many other similarly stimulating topics are discussed in the twelve chapters of this volume that would be of interest to a wide spectrum of readers including ecological and agricultural scientists as well as agricultural practitioners. However, information presented in this book leads this reader to conclude that a large gap still exists between theoretical ecology and its application to problem solving in the realm of sustainable agriculture. Disciplines within ecology are at different stages of preparedness to date, and little cooperative effort at the appropriate level is ongoing. There does, however, appear to be a consensus that ecologists must begin to rally around a new, transdisciplinary "agrolandscape ecology" where teams including farmers, agricultural scientists, and legislators work together to develop and manage sustainable agriculture. Building teams which include ecologists, agronomists, and soil scientists is essential. Ecologists must not overlook the many contributions that soil science and agronomy have already made to the arena of sustainable agriculture. Soil scientists and agronomists must be linked with ecologists to maximize both the profitability and environmental sustainability of agriculture. Neither ecologists nor agriculturalists alone can bring about the desired end. Integrating ecological principles and agriculture in new ways can lead to a sustainable agriculture and the development of sound environmental policies.

G. A. Peterson
Professor of Agronomy
Colorado State University
Fort Collins, CO

CONFERENCE PROCEEDINGS

Integrating Sustainable Agriculture, Ecology, and Environmental Policy

Richard K. Olson

SUMMARY. Current agricultural practices are contributing to environmental degradation, which also threatens the sustainability of agricultural production. Ecology has the potential to contribute significantly to the development of a sustainable and environmentally sound agriculture. However, the results of a conference organized by the U.S. Environmental Protection Agency suggest that only a small part of this potential has been realized. Successful translation of ecological research results into agricultural management decisions will require a multidisciplinary approach. Ecologists may be most effective in influencing agricultural practices by working as members of multidisciplinary research teams that include farmers and on-farm research.

Richard K. Olson is Project Scientist, Sustainable Agriculture Project, ManTech Environmental Technology, Inc., USEPA Environmental Research Laboratory, 200 SW 35th Street, Corvallis, OR 97333.

This article has been prepared with funding from the U.S. Environmental Protection Agency. It was prepared at the EPA Environmental Research Laboratory in Corvallis, OR, through contract #68-C8-0006 to ManTech Environmental Technology, Inc. It has been subjected to the Agency's peer and administrative review and approved for publication.

INTRODUCTION

> ***Land that is in human use must be lovingly used; it requires intimate knowledge, attention, and care.***
>
> —Wendell Berry

Most of the land in the United States is in human use. In the 48 conterminous states, 23 percent of the land area is used as cropland (U.S. DOC 1990), 30 percent is rangeland (USDA 1981), and 25 percent is classified as commercial timberland (USDA 1982). Less than 2 percent is formally protected as wilderness (Reed 1989). We depend upon managed lands for commodities and as sites for cities, roads, and the other components of human infrastructure. We also depend on managed lands for noneconomic ecosystem services that are just as critical to our well-being: air and water purification, climate modification, habitat for other species, and aesthetics. We have to ensure that managed lands perform these multiple services because managed lands are virtually all that we have. What remains of the original landscape is a "semi-natural matrix" (Roberts 1988) within which humans and other species must survive.

However, the condition of this matrix and the ability of managed ecosystems in the United States to provide services is deteriorating.

- Approximately one-half of all wetlands outside of Alaska have been lost since settlement, with 87 percent of the losses due to conversion to agriculture (Dahl 1990).
- Soil erosion from croplands is estimated to be about 3 billion tons per year (NRC 1989), with one-fifth of cropland subject to serious damage. Erosion also degrades forest lands; 435 million tons of soil eroded from non-Federal forest lands in 1977 (USDA 1981).
- Sixty-eight percent of Bureau of Land Management lands in the West are reported to be in unsatisfactory condition due to overgrazing (Frazier 1990).
- Eighty to ninety percent of old-growth forests in the Pacific Northwest have been eliminated (Henderson et al. 1989).
- Approximately 1.5 million acres of U.S. farmland are converted annually to other uses such as development (AFT 1991).
- Approximately two-thirds of California's native fish taxa have declining, threatened, or endangered populations (Moyle and Williams 1990).
- Salinization negatively affects crop yields on up to 4 million acres of irrigated cropland in California's Central Valley (Shafroth 1989).

- One-fourth of counties surveyed by the U.S. Geological Survey showed elevated nitrate-nitrogen concentrations in at least 25 percent of the wells tested (NRC 1989).

A much longer list could be developed. Reversing these trends will be extremely complex, and will require a great deal of "knowledge, attention, and care." The necessary knowledge comes from virtually every discipline including sociology, agronomy, forestry, political science, economics, and ecology. Knowledge of ecosystem structures and functions and how these contribute to system resilience and resistance to stresses is increasingly being used to improve management practices (e.g., Perry and Maghembe 1989, Gliessman 1990, Raven 1990). Most recently, the Ecological Society of America's Committee for a Research Agenda for the 1990's recommended sustainable ecological systems as one of three research priorities in a Sustainable Biosphere Initiative. Research would "focus on understanding the underlying ecological processes in natural and human-dominated systems in order to prescribe restoration and management strategies that would enhance the sustainability of the Earth's ecological systems" (Lubchenco et al. 1991).

CONFERENCE

The role of ecological research in supporting development of one aspect of sustainable use, sustainable agriculture, was evaluated at a conference organized by the U.S. Environmental Protection Agency (EPA) in Arlington, Virginia, July 22-23, 1991. Agriculture contributes to many of the environmental problems for which EPA is seeking solutions; for example, nonpoint source pollution, contamination of groundwater, air toxics, and loss of biodiversity. EPA's new emphasis on prevention rather than mitigation of environmental problems makes the potential environmental benefits of sustainable agriculture attractive to the Agency. EPA's growing expertise in ecology, combined with a recognition of the importance of ecological knowledge as a foundation for sustainable agriculture, led to the conference theme.

The objectives of the conference included examining the socioeconomic and political context within which sustainable agriculture must develop, discussing the application of ecological knowledge to developing a sustainable agriculture within this context, and identifying research priorities. These issues were addressed by a series of papers that were presented at the conference and are printed in this volume.

CONFERENCE RESULTS

The conference was attended by scientists and administrators from EPA and USDA, academics from a variety of universities and disciplines, state and local government officials, farmers, private consultants, and others. The diversity of backgrounds and perspectives led to a corresponding diversity of ideas and opinions. There were, however, several themes that emerged from the discussions.

Foremost was the feeling that, while ecology has much to offer in developing sustainable agriculture, its promise has been largely untapped. Although the logic of ecological science as a foundation for sustainable agriculture has been clearly stated (e.g., Jackson and Piper 1989, Power and Kareiva 1990), examples of the application of ecology to designing sustainable management systems are not abundant. Integrated Pest Management (Andow and Rosset 1990), conservation strategies for wildlife in agricultural regions (Fry 1990), and spatial patterning of cultivated and uncultivated strips (Kemp and Barrett 1989) were among the examples discussed at the conference. However, much of the work in agroecology involves experiments with agricultural systems to elucidate ecological processes or the use of ecological knowledge to explain observed functions of particular systems. Both are important, but are not the same as using ecological knowledge to design a sustainable system.

One reason for this unrealized potential is the complexity of applied agroecology relative to ecological studies of natural systems. Successfully applying ecology to agriculture requires attention to the socioeconomic context within which sustainable agriculture functions. For example, an ecologically based management scheme that bankrupts the farmer is not sustainable. Research results with which farmers feel no affinity or ownership will probably not be implemented. Sustainable agriculture is especially multidisciplinary, and ecologists often do not wish to work in this environment.

The disciplinary complexity of sustainable agriculture was illustrated by the responses of a panel of speakers when asked: "What one indicator of agricultural sustainability would each panelist recommend?" Answers included:

- ratio of soil erosion to soil formation
- trend in average age of farmers
- nematode demographics
- connectivity of landscape elements
- net farm income
- water quality.

The panelists emphasized that no single indicator would suffice as a measure of sustainability–more complex indices need to be developed. Identifying and testing indicators of ecological condition to serve as components of these broader indices needs to be a research priority of agroecology.

Another general research priority identified by the participants was long-term (more than 5 years) and large-scale (watershed/landscape) studies. Most current research in sustainable agriculture focuses on plots, fields, or farms, and long-term funding is rare. However, most of the environmental issues associated with agriculture are manifested at large scales. Measurable changes in water quality, soil quality, biodiversity, and other environmental indicators in response to new agricultural practices may take many years to occur. The only way to fully test and validate these relationships is to use the necessary time and space. Full-scale demonstrations are also critical tools in gaining public and governmental acceptance of new approaches.

CONCLUSIONS

In spite of the challenges, agroecology is growing rapidly, as evidenced by the increasing volume of literature in the field (e.g., Gliessman 1990; Carroll et al. 1990). Agroecology research programs such as those at the Land Institute in Kansas (Piper and Gernes 1989), and the Kellogg Biological Station in Michigan (Franklin et al. 1990) are making important contributions. However, translation of agroecology research results into sustainable management practices will not occur automatically. Ecologists must take active steps to make this happen, including participation as members of multidisciplinary research teams that include farmers and conduct on-farm research (Francis et al. 1990; Flora, this volume).

Ecology offers no magic solutions to hard problems. A more ecologically based agriculture could increase production efficiency and decrease environmental impacts, but resource constraints will still set an upper limit to sustainable production. The hard choices regarding population control, energy conservation, land use, and the other components of a sustainable society must still be made. Ecology can help to guide those decisions, but it will not provide the means to avoid them.

REFERENCES

American Farmland Trust. 1991. The battle plan. *Am. Farmland*, Spring: 5-6.

Andow, D. A., and P. M. Rosset. 1990. Integrated pest management. In *Agroecology*, ed. C. R. Carroll, J. H. Vandermeer, and P. M. Rosset, 413-39. New York: McGraw-Hill.

Carroll, C. R., J. H. Vandermeer, and P. M. Rosset, eds. 1990. *Agroecology*. New York: McGraw-Hill. 641 pp.

Dahl, T. E. 1990. *Wetlands losses in the United States 1780's to 1980's*. Washington, D.C.: U.S. Department of the Interior, Fish and Wildlife Service. 21 pp.

Francis, C., J. King, J. DeWitt, J. Bushnell, and L. Lucas. 1990. Participatory strategies for information exchange. *Am. J. Altern. Agric.* 5(4):153-60.

Franklin, J. F., C. S. Bledsoe, and J. T. Callahan. 1990. Contributions of the long-term ecological research program. *Bioscience* 40(7):509-23.

Frazier, D. 1990. Grazing helps land, livestock industry says. *Rocky Mountain News*, 14 January 1990.

Fry, G. L. A. 1990. Conservation in agricultural ecosystems. In *The scientific management of temperate communities for conservation*, ed. I. F. Spellerberg, F. B. Goldsmith, and M. G. Morris, 415-43. Oxford, UK: Blackwell Scientific Publications.

Gliessman, S. R., ed. 1990. *Agroecology: Researching the ecological basis for sustainable agriculture*, New York: Springer-Verlag. 512 pp.

Henderson, S., R. K. Olson, and R. F. Noss. 1989. Current and potential threats to biodiversity in forests of the lower Pacific Coastal States. In *Transactions: Symposium on the effects of air pollution on western forests, June 29-30, Anaheim, CA*, ed. R. K. Olson and A. S. Lefohn, 325-36. Pittsburgh, PA: Air & Waste Management Association.

Jackson, W., and J. Piper. 1989. The necessary marriage between ecology and agriculture. *Ecology* 70(6):1591-93.

Kemp, J. C., and G. W. Barrett. 1989. Spatial patterning: Impact of uncultivated corridors on arthropod populations within soybean agroecosystems. *Ecology* 70(1):114-28.

Lubchenco, J., and fifteen co-authors. 1991. The Sustainable Biosphere Initiative: An ecological research agenda. *Ecology* 72(2):371-412.

Moyle, P. B., and J. E. Williams. 1990. Biodiversity loss in the temperate zone: Decline of the native fish fauna of California. *Conserv. Biol.* 4(3):275-84.

National Research Council. 1989. *Alternative agriculture*. Washington, D.C.: National Academy Press. 448 pp.

Perry, D. A., and J. Maghembe. 1989. Ecosystem concepts and current trends in forest management: Time for reappraisal. *For. Ecol. Manage.* 26:123-40.

Piper, J. K., and M. A. Gernes. 1989. Vegetation dynamics of tallgrass prairie and their implications for sustainable agriculture. In *Prairie pioneers: Ecology, history, and culture*, ed. T. B. Bragg, 9-14. Lincoln, NE: University of Nebraska Press.

Power, A. G., and P. Kareiva. 1990. Herbivorous insects in agroecosystems. In *Agroecology*, ed. C. R. Carroll, J. H. Vandermeer, and P. M. Rosset, 301-27. New York: McGraw-Hill.

Raven, P. H. 1990. The politics of preserving biodiversity. *Bioscience* 40(10):769-74.

Reed, P. C. 1989. The National Wilderness Preservation System: The first 23 years and beyond. In *Wilderness benchmark 1988: Proceedings of the National Wil-*

derness Colloquium, Tampa, Florida, January 13-14, 1988, 2-20. General Technical Report SE-51. U.S. Department of Agriculture, Forest Service, Southeastern Forest Experiment Station.

Roberts, L. 1988. Hard choices ahead on biodiversity. *Science* 241:1759-61.

Shafroth, M. 1989. Changes and challenges in the Central Valley. *Am. Farmland*, Summer: 1-3.

U.S. Department of Agriculture. 1981. *An assessment of the forest and range land situation in the United States*. Forest Resource Report no. 22. Washington, D.C.: USDA, Forest Service.

U.S. Department of Agriculture. 1982. *An analysis of the timber situation in the United States, 1952-2030*. Forest Resource Report no. 23. Washington, D.C.: USDA, Forest Service.

U.S. Department of Commerce, Bureau of the Census. 1990. *1987 Census of agriculture*. Washington, D.C.: U.S. Government Printing Office.

The Future Context of Sustainable Agriculture: Planning for Uncertainty

Richard K. Olson

SUMMARY. Agricultural management systems function in an environmental, social, and economic context. The future context of agriculture in the United States will differ from the present in ways that cannot be predicted with certainty, but will likely challenge the sustainability of current systems. Research on sustainable agricultural systems must consider this uncertainty, and policy decisions must take uncertainty into account. Approaches for dealing with uncertainty include planning for extreme events, adopting policies with multiple benefits, improving predictive capabilities, and educating decision makers and citizens on the implications of future uncertainties.

INTRODUCTION

One of the themes of this volume is the use of ecological knowledge in developing a sustainable agriculture. Developing a sustainable agriculture is essentially a process of altering, where necessary, current management practices. Unless ecological knowledge is used, directly or indirectly, in the formulation and testing of management practices, it will not contribute to meeting the goal of a sustainable agriculture.

Richard K. Olson is Project Scientist, Sustainable Agriculture Project, ManTech Environmental Technology, Inc. USEPA Environmental Research Laboratory, 200 SW 35th Street, Corvallis, OH 97333.

This article has been prepared with funding from the U.S. Environmental Protection Agency. It was prepared at the EPA Environmental Research Laboratory in Corvallis, OR, through contract #68-C8-0006 to ManTech Environmental Technology, Inc. It has been subjected to the Agency's peer and administrative review and approved for publication.

This focus on applied research complicates the work of ecologists working with agriculture. Agricultural management systems function within an environmental and social context. They succeed or fail in the "real world," and researchers therefore need to consider this context in planning research and assessing its results.

Key variables that must be considered in developing an agricultural management system are listed in Table 1. Successful management systems must deal appropriately with the full suite of contextual variables, including interactions between variables. Individual variables will differ from place to place, with changes occurring at scales ranging from regional (e.g., climate) to farm (e.g., slope of the land) to field (e.g., soil type). As a result, management systems appropriate for one location may not be appropriate for other locations.

Researchers and managers are well aware of this issue of site specificity (Blake 1990, Walters et al. 1990). Research results are commonly evaluated for their generality or extent to which they can be extrapolated to other sites or regions. Sustainable agriculture may be even more site-specific than conventional agriculture because of its greater reliance on information and lesser reliance on chemical and energy inputs (Bunch 1990).

Less thought is given by researchers to determining the applicability of research results to future contexts than is given to the issue of site specificity (i.e., temporal rather than spatial extrapolation). Every variable listed in Table 1 can change over time, and some can change over fairly short time scales. Failure to anticipate these changes can mean that management systems that work *now* for a specific location may not work in the *future* at that

Table 1. Examples of variables included in the context of an agricultural system.

Environmental	Socioeconomic
Climate	Human demographics
Topography	Government agricultural policies
Soil types and condition	Consumer preferences
Native biota	Transportation infrastructure
Air and water pollution	Commodity prices
Crop and livestock gene pools	Consumer preferences
Ground and surface water supplies	Agricultural technology
Potential erosion hazard	Cost of inputs

same location. For example, many midwestern farmers who thrived under the economic conditions of the 1970s failed when the economic environment changed in the 1980s (NRC 1989).

Other components of the agricultural context are as critical as the economic climate, and as difficult to predict. Because the future cannot be accurately predicted, sustainable agriculture must plan for uncertainty. Researchers need to consider how the context in which their results will be applied may differ from the context in which the research was conducted. They must then evaluate the implications of future uncertainty for the usefulness of their work.

This paper illustrates through examples the types of future uncertainties faced by U.S. agriculture. A general strategy for dealing with this uncertainty is described, followed by specific recommendations for research and policy.

FUTURE TRENDS AND UNCERTAINTIES

Uncertainties in predicting the future context of sustainable agriculture in the United States can be illustrated through discussion of potential trends in three key environmental and social variables: human population size, climate, and environmental condition of agricultural landscapes. Each of these variables has an important influence on the success of any farming system.

The size of the human population is the most important factor shaping the future context of agriculture. Population size sets the lower bounds for the levels of food and fiber production that must be sustained by agroecosystems. Human populations also compete with agriculture for land and other resources (e.g., water in California). Population growth in other countries may influence U.S. agriculture through processes such as greenhouse gas emissions, world trade, competition for oil, and immigration.

The importance of climate to agricultural systems is obvious. Regarding environmental quality–soils, water, and other ecosystem components form the base of agricultural productivity. Their degradation has a direct impact on agricultural production as well as other essential ecosystem functions.

Human Population

The current world population is 5.4 billion, with an annual rate of increase of 1.7% (PRB 1991). This means a population increase of 92 million in 1991 alone and, if this rate is maintained, world population will double in 40 years. However, many social and environmental factors influence population growth, and actual trends cannot be predicted with confidence. Three alternative population scenarios based on different assumptions for demographic variables (UN 1991) are shown in Figure 1. Estimates of world population in the year 2025 range from 7.6 billion to 9.4 billion.

Although 95% of the projected world population increases will occur in Third World countries (UN 1991), the U.S. population also is growing rapidly. The estimated U.S. population on February 1, 1991, was 251,593,000, up 2,742,000 over the previous year (U.S. BOC 1991a), for an annual rate of increase of 1.1%. Continued growth at this rate would double the U.S. population in 62 years.

FIGURE 1. World population projections: medium, high, and low variants. Data from United Nations (1991).

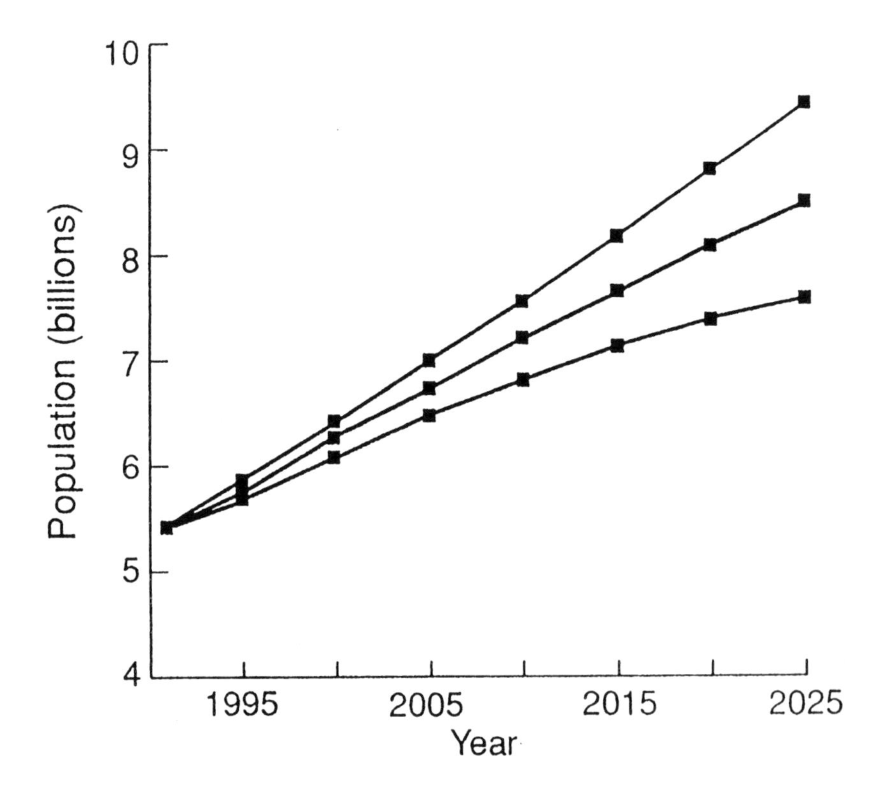

Actual population growth in the United States will depend primarily on the Total Fertility Rate (TFR) and levels of immigration. The TFR represents the average number of surviving children a woman will have during her lifetime. Figure 2 shows four projected scenarios based on different values for TFR and immigration. The TFR during the 1980s averaged 1.8 (Bouvier 1989), while net immigration (legal and illegal) is estimated to have been approximately 600,000 per year (Weber 1988). This scenario lies between curves A and B. However, the TFR has risen to 2.0 (U.S. BOC 1991b) and recent legislation could increase legal immigration alone to more than 1 million annually (Anon. 1991). When illegal immigration is also considered, the predicted future population for the United States more closely

FIGURE 2. United States population projections for different combinations of Total Fertility Rate (TFR) and net annual immigration (NAI). A: TFR = 1.7, NAI = 0.5 million; B: TFR = 1.9, NAI = 1.1 million; C: TFR = 1.9, NAI = 1.6 million, D: TFR = 2.1, NAI = 2.0 million. From PEB (1987) with permission.

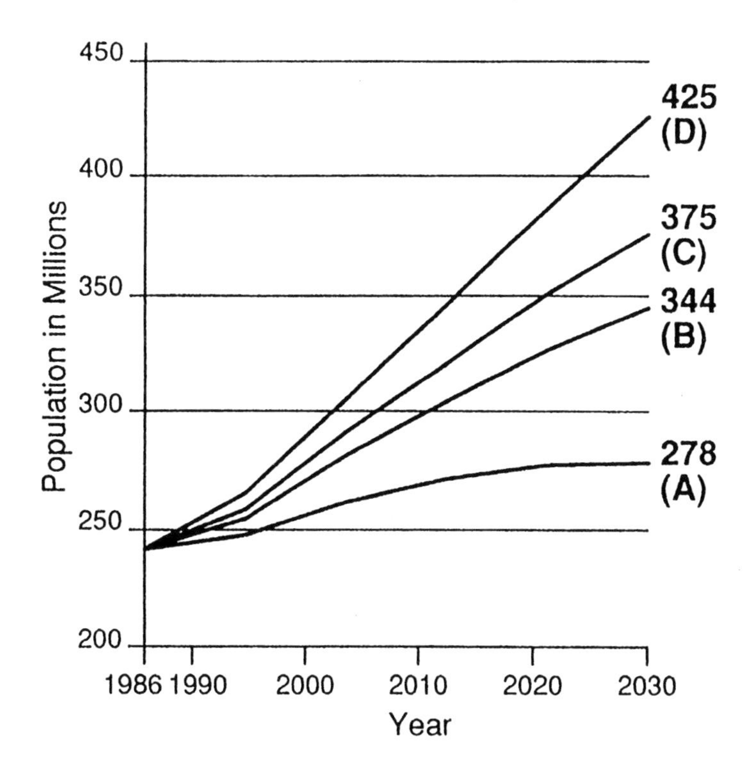

resembles curve C. These rapid changes in TFR and immigration demonstrate the volatility of population projections.

Climate

It is uncertain whether the United States will maintain its current climate in the near future. Atmospheric concentrations of greenhouse gases have increased since the mid-18th century by approximately 50% (CO_2 equivalent), creating the potential for global warming through an enhanced greenhouse effect (Houghton et al. 1990). Global mean surface air temperature has increased by 0.3°C to 0.6°C over the last 100 years, and is now the warmest on record. However, the increase is not yet great enough to be distinguished from natural variability (Wigley and Barnett 1990).

Predictions of future global temperature (Figure 3) are uncertain, in part because atmospheric processes are not completely understood (Houghton et al. 1990). Estimates of future emissions of greenhouse gases upon which

FIGURE 3. Simulation of the increase in global mean temperature from 1850-1990 due to observed increases in greenhouse gases, and predictions of the rise between 1990 and 2100 resulting from the Business-as-Usual emissions scenario. High, best, and low estimates are shown. From Houghton et al. (1990) with permission.

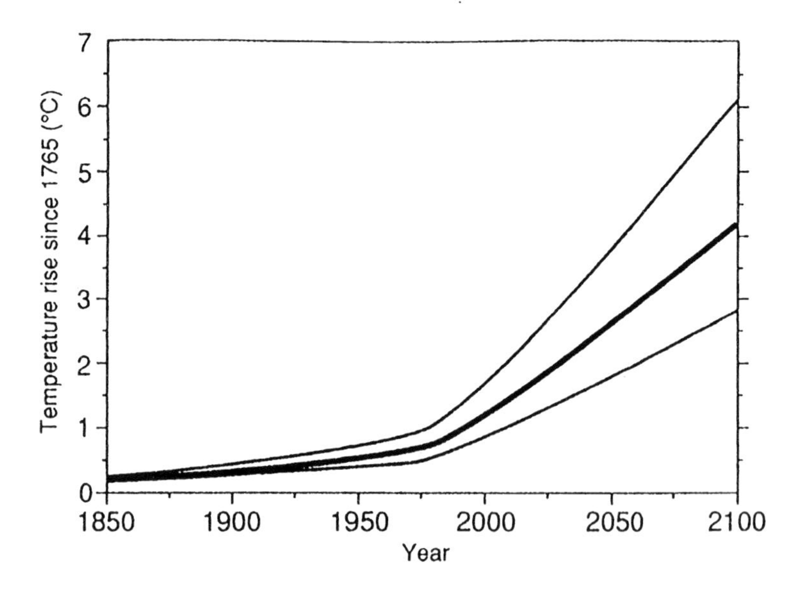

temperature predictions are based also vary (Houghton et al. 1990). Implementation of policies such as the Montreal Protocol to reduce CFC production (Stetson 1990) or shifts to alternative energy sources could decrease emissions. Conversely, burgeoning world populations coupled with attempts to raise Third World living standards could increase emissions.

The exact effects of a global temperature increase on the climate of the United States are uncertain. Changes in precipitation patterns need to be considered along with temperature, and the magnitude and even the direction of change is likely to vary regionally (Houghton et al. 1990). Assuming a global mean warming of 1.8 °C by the year 2030, the predictions by three high-resolution climate models for summer in central North America are: temperature increases by 2-3 °C; precipitation decreases by 5-10%; and soil moisture decreases 15-20% (Mitchell et al. 1990). These predictions are considered highly uncertain. They also change by -30% to +50% if different assumptions of atmospheric sensitivity are used in the models (Mitchell et al. 1990).

Environmental Condition of Agricultural Landscapes

We currently lack proven indicators of environmental condition for agricultural landscapes and the data to make conclusive statements about regional environmental condition in the United States (Neher, this volume). However, statistics on three potential indicators are illuminating:

- An estimated 1.5 million acres of farmland are converted to nonagricultural uses each year (AFT 1991). Much of this loss is irreversible and can have as a secondary consequence the fragmentation and reduction in function of the remaining agricultural lands.
- Average annual sheet and rill erosion for U.S. cropland in 1987 was 3.8 tons per acre (Lee 1990). Wind erosion is significant in some regions, and overall approximately one-fifth of U.S. cropland is subject to serious erosion damage (NRC 1989).
- Twenty percent of the native flora of the United States is considered by the Nature Conservancy to be at risk (McMahan 1990). Agriculture is an important contributor to this problem; however, its contribution relative to other landuses is not known.

Future trends in these indicators are even more difficult to predict. Many localities have instituted land use planning to reduce farmland loss; however, zoning can be overturned as development pressures mount. For example, Oregon is currently considering legislation to remove millions of acres from Exclusive Farmland Use designation.

Average annual sheet and rill erosion on U.S. cropland declined almost 12% between 1982 and 1987, due primarily to Federal acreage reduction programs and conservation management improvements (Lee 1990). Economic shifts or changes in Federal policies, however, could reduce participation in these programs (see Doering, this volume). One of the key Federal laws to protect biodiversity, the Endangered Species Act, faces legislative pressure to weaken it.

In spite of all the uncertainties, some statements can be made with confidence about the future context of agriculture. Human demands for food, fiber, and other natural resources will be much greater, while the natural resource base will be significantly diminished relative to the present. The fact that we do not know exactly how much more crowded, degraded, or warmer the United States will be should not prevent us from acknowledging general trends.

PREDICTING EFFECTS

It is not changes per se in the context of agriculture that are of interest, but their potential effects on agricultural sustainability. Unfortunately, predicting effects is doubly difficult; not only do we lack good predictions of the future context, but we also have a poor understanding of the relationships between changes in context and changes in agricultural sustainability.

Uncertainty in predicting effects is heightened by the fact that the future magnitudes of important stressors may be outside the range of our experience, thus forcing us to extrapolate beyond the bounds of our data. The earth and the United States have never had as many people as now; agricultural systems have never faced a global climate as warm as that predicted for the next century; and society has never dealt with a natural resource base as degraded as now. The potential for nonlinear responses makes accurate projections even more difficult.

The debate over population growth in the United States serves as a good example of the uncertainty inherent in predicting the effects of stressors on the sustainability of agricultural and social systems. One of the key questions is defining carrying capacity, which in turn is a function of population size and per capita resource consumption. At one extreme of the debate, Pimentel and Pimentel (1990) estimate the carrying capacity of the United States, at current rates of per capita resource utilization, to be 40-100 million people. In contrast, Simon (1989), and Simon and Kahn (1984), suggest that technological advances will support virtually unlimited U.S. population growth.

Clearly, at least one of these estimates is wrong. And while this may be an extreme example, policy-makers frequently must reconcile divergent views. Given the potential for disastrous policy choices, what is the best strategy for dealing with uncertainty?

A STRATEGY FOR DEALING WITH UNCERTAINTY

A prudent strategy seeks to minimize the chances of a disastrous outcome (Costanza 1990). A simplified "pay-off matrix" (Figure 4) suggests that policy decisions for achieving a sustainable agriculture in the face of large uncertainties should be based on some of the more pessimistic predictions of the future context. Policies based on the assumptions that population growth or climate change are not serious threats to sustainability, or that technological advances will provide fixes, may lead to disaster if these assumptions are incorrect. Alternatively, policies that plan for the worst may risk mainly the loss of some short-term economic gains if an optimistic view of the state-of-the-world proves to be correct.

Within the bounds of a prudent approach, some general recommendations for strategies to deal with uncertainty are listed below. They are divided into recommendations for research and for policy.

Research

- *Research must be multidisciplinary.* Success in the "real world" requires attention to the whole context. Research groups need to include farmers and other nonscientists.
- *Emphasize applied research.* Each year that passes without action makes it more difficult to reverse the trends of environmental stressors and degradation. Estimates of the time remaining to achieve critical environmental goals include ten years for protecting global biodiversity (Roberts 1988) and forty years for developing a sustainable society (Brown et al. 1990). The social and environmental trends discussed in this paper are compelling reasons why society can no longer afford the luxury of funding research whose results simply end up on a dusty bookshelf.
- *Emphasize analysis of existing data.* Existing datasets offer cost and time advantages over collection of new data, but often are overlooked. A first step would be a comprehensive listing and evaluation of existing datasets to determine their utility in addressing research questions

concerning sustainable agriculture. New datasets should include proper documentation so that other researchers can use them.

- *Emphasize research to improve predictive capabilities.* Policy decisions require better predictions of the effects of alternative management scenarios. Improved mechanistic models at a variety of scales is a critical requirement.
- *Researchers must spend time on education and technology transfer.* Researchers' predictions and recommendations will be accepted only if policymakers and the public understand the rationale behind them.

Policy

- *Policies must be multidisciplinary.* No agricultural systems are sustainable if issues such as population growth, global climate change, economic trends, and environmental degradation are not addressed.
- *Actions should not require absolute certainty.* Given the long lag times inherent in reversing global environmental trends or preparing for them, actions must be initiated before trends are fully manifest.
- *Policies should take extreme events into account.* Droughts, disease outbreaks, typhoons, and other unexpected disasters happen. Planners should not assume that harvests will be optimal or even average every year.
- *Adopt policies with multiple benefits.* This amounts to hedging your bets. For example, increased taxes on gasoline would decrease use and CO_2 emissions. They also could reduce dependence on foreign energy sources, reduce trade and fiscal deficits, and help to conserve a scarce resource. Even if climate change turns out not to be a problem, much would have been gained.

CONCLUSIONS

The future context of agriculture is very uncertain; some might say bleak. However, there are no physical reasons why a sustainable agriculture and society cannot be developed. The knowledge necessary to design sustainable agricultural systems exists or is being developed. Sufficient resources still exist to provide a foundation for a sustainable society, and general approaches such as those outlined in this paper can help to attain that goal. It remains to be seen, however, whether society and policymakers are willing to make the hard choices and sacrifices required to attain sustainability. That is the greatest uncertainty.

FIGURE 4. Payoff matrix for policy optimism vs. pessimism. Based on Costanza (1990) with permission. The four combinations of types of policy and actual state of the world give payoffs to society ranging from disaster to high. See text for details.

Adopted Policy	Real State of the World: Optimists Right	Real State of the World: Pessimists Right
Optimist Policy	**High**	**Disaster**
Pessimist Policy	**Moderate**	**Tolerable**

REFERENCES

American Farmland Trust. 1991. The battle plan. *Am. Farmland*, Spring: 5-6.

Anonymous. 1991. How the 1990 immigration law adversely affects the United States. *Balance Data*, no. 28. Washington, D.C.: Population-Environment Balance.

Blake, R. O. 1990. Nine questions most often asked about global agricultural sustainability. *J. Soil Water Conserv.* 45(4):460-62.

Bouvier, L. F. 1989. How to get there from here: The demographic route to optimal population size. *The NPG Forum*, no. 2. Washington, D.C.: Negative Population Growth.

Brown, L. R., C. Flavin, and S. Postel. 1990. Earth Day 2030. *World Watch* 3(2):12-21.

Bunch, R. 1990. The meaning and benefits of partnership in agricultural research: Past successes–future potentials. *Am. J. Altern. Agric.* 5(4):147-50.

Costanza, R. 1990. Balancing humans in the biosphere: Escaping the overpopulation trap. *The NPG Forum*, no. 8. Washington, D.C.: Negative Population Growth. 6 pp.

Houghton, J. T., G. J. Jenkins, and J. J. Ephraums, eds. 1990. *Climate change: The intergovernmental panel on climate change scientific assessment.* Cambridge, UK: Cambridge University Press.

Lee, L. K. 1990. The dynamics of declining soil erosion rates. *J. Soil Water Conserv.* 45(6):622-24.

McMahan, L. R. 1990. Propagation and reintroduction of imperiled plants, and the role of botanical gardens and arboreta. *Endangered Species Update* 8(1):4-7.

Mitchell, J. F. B., S. Manabe, V. Meleshko, and T. Tokioka. 1990. Equilibrium climate change–and its implications for the future. In *Climate change: The intergovernmental panel on climate change scientific assessment*, ed. J. T. Houghton, G. J. Jenkins, and J. J. Ephraums, 131-75. Cambridge, UK: Cambridge University Press.

National Research Council. 1989. *Alternative agriculture*. Washington, D.C.: National Academy Press.

Pimentel, D., and M. Pimentel. 1990. Land, energy, and water: The constraints governing ideal U.S. population size. *The NPG Forum*, no. 3. Washington, D.C.: Negative Population Growth.

Population-Environment Balance. 1987. U.S. population projections from 1986-2030. *Balance Report*, no. 53. Washington, D.C.

Population Reference Bureau. 1991. *World population data sheet*. Washington, D.C.

Roberts, L. 1988. Hard choices ahead on biodiversity. *Science* 241:1759-61.

Simon, J. L. 1989. *The economic consequences of immigration*. New York: Basil Blackwell. 402 pp.

Simon, J. L., and H. Kahn, eds. 1984. *The resourceful earth*. New York: Basil Blackwell. 585 pp.

Stetson, M. 1990. Who'll pay to protect the ozone layer? *World Watch* 3(4):36-37.

United Nations. 1991. *World population prospects 1990*. Population Studies, no. 120. New York: United Nations Population Division.

U.S. Bureau of the Census. 1991a. Population estimates and projections. *Current Population Reports*, Series P-25, no. 1072. Washington, D.C.: U.S. Government Printing Office.

U.S. Bureau of the Census. 1991b. *Monthly Vital Statistics Report* 39(9). Washington, D.C.: U.S. Government Printing Office.

Walters, D. T., D. A. Mortensen, C. A. Francis, R. W. Elmore, and J. W. King. 1990. Specificity: The context of research for sustainability. *J. Soil Water Conserv.* 45:55-57.

Weber, S. 1988. *USA by numbers: A statistical portrait of the United States*. Washington, D.C.: Zero Population Growth.

Wigley, T. M. L., and T. P. Barnett. 1990. Detection of the greenhouse effect in the observations. In *Climate change: The intergovernmental panel on climate change scientific assessment*, ed. J. T. Houghton, G. J. Jenkins, and J. J. Ephraums, 239-55. Cambridge, UK: Cambridge University Press.

Federal Policies as Incentives or Disincentives to Ecologically Sustainable Agricultural Systems

Otto Doering

SUMMARY. Traditionally, American farm policies have dealt with the cost of food, the supply of commodities, and incomes of farmers. Concerns about environmental impacts have been peripheral. The resource concerns that led to institutions like the Soil Conservation Service were linked closely with income and other program goals. The programs we have today were not designed to address the new resource uses and sustainability goals our society now desires.
Relative resource prices have given strong signals to farmers to use more of some resources than others. Inexpensive petrochemicals in contrast to more expensive land and labor have encouraged the substitution of these less expensive inputs for the more expensive ones. There has been a strong trend of such substitution since the Second World War.

A whole new outlook toward farm size, cropping systems, and relative resource use has evolved since the coming of the tractor. The mechanization of agriculture allowed modern chemicals and responsive varieties to have tremendous impact that further changed the face of agriculture away from more traditional mixed enterprises and cropping systems.

Analysis shows that changes in existing farm programs, or even the elimination of these programs, will neither result in very large relative or absolute changes in input use, nor in basic changes in the way farmers farm. A number of nonfarm policies and events have been more important to farm structure, input use, and crop mix decisions than have so-called "farm" programs.

Otto Doering is Professor of Agricultural Economics and Public Policy Specialist at Purdue University, West Lafayette, IN 47907-1145.

Existing farm programs were neither designed for, nor were effective in, changing the way farmers farm. Therefore, new types of policies will have to be created–policies that might target specific concerns, reassign property rights, or give very different direct incentives in order for farmers to change the way they farm.

INTRODUCTION

Many papers and proposals have been written over the last year as part of the Farm Bill debate. These analyzed or recommended specific policy changes to achieve what the authors believed would be more environmentally benign or sustainable agriculture (e.g., Reichelderfer and Phipps 1988). That is not the intent here. Instead, our intent is to provide perspective on farm policy's impact on the way farmers farm. A broader perspective is presented along with the likely consequences of more sweeping policy actions than those ultimately included in the 1990 Farm Bill. The critical question is whether Federal "farm" policy is the most effective vehicle for changing agricultural practices and impacts.

No specific definition of sustainable agriculture is embraced here. However, the implicit notion is that sustainable agriculture is a structurally different system from today's system. Sustainable agriculture involves less use of off-farm inputs while introducing new management and cropping systems that better utilize on-farm resources. These systems allow continued farming over the long term with sustained productivity and minimal stress on the environment.

Until recently, Federal policies toward agriculture have not been concerned with environmental issues. Rather, they have reflected the goals of increased farm income and of low cost food for the nation. The food system has provided a continually declining relative cost of food (Manchester 1991). Federal policies have had other goals such as improving rural conditions (Bailey et al. 1909; Tolley 1940), increasing rural income, improving the technical efficiency of farming operations, and conserving natural resources as part of the base of agricultural productivity. At times, policies to achieve one goal have conflicted with another goal. Also, as is true with other industries, the market does not give signals to agricultural producers that reflect the cost of environmental damage; this then provides a rationale for government intervention.

Future policies will have to be concerned with the environmental impacts of agriculture. Federal policy will attempt to deal with this in ways that do as little economic damage to the agricultural sector as possible. The Federal policy interest in sustainable agriculture relates more to its hoped-for environmentally benign characteristics than to the inherent character of its internal management, farm resource use, or social characteristics. As a result, one may view the latter items as consequences of sustainable systems rather than as specific policy goals. The question should be "How might agriculture be more environmentally benign?" not "How might it be more sustainable?" The answers may or may not coincide.

Our discussion will begin with a look at current systems and the forces that drive these systems. Then the following topics will be investigated: farming characteristics that affect sustainability, different Federal policies that might change the way farmers farm, some of the changes under the 1990 Farm Bill, incentives to change which are on the horizon, and where Federal policy might be headed.

CURRENT AGRICULTURAL SYSTEMS

Farmers are farming the way they are today for a wide variety of reasons. These include the particular resource endowment a farmer faces, the strength of demand (prices) for different products he or she might produce, the different alternative technologies and inputs available for use in production, and the relative costs of different technologies and inputs a farmer might utilize in the production of different products. Reasons also include the various regulations and incentives under farm and nonfarm government policies as well as the impact of the broader national economic and political environment. Policies toward agriculture and farmers are only one part of the farmer's decision framework. For farm policies to dominate a farmer's decisions, they must predominate over the other factors influencing a farmer's choices.

There have been important changes in the relative value of farm inputs that have influenced farm production decisions. Land, labor, energy, and farm machinery became increasingly expensive relative to pesticides and fertilizer from 1950 to 1987, as illustrated in Table 1 (Conway et al. 1991; U.S. Department of Agriculture 1985, 1988). Table 2 shows the changes in utilization of some of these inputs over time (Hallberg 1988; U.S. Department of Agriculture 1991). Aside from the impacts of government programs, the utilization of land has changed the least. There have been some important regional shifts in the specific applications of agricultural land, but the crop base has been relatively stable.

Certainly, government programs that restricted acreage and increased commodity prices gave farmers incentives to intensify production and use

Table 1. Five-year average input prices (indices with 1982 = 100).

Years	Land	Pesticides	Fertilizer	Total Labor	Gasoline	Farm Machinery
1948-52	10	71	48	23	18	14
1953-57	12	61	46	26	20	21
1958-62	15	56	43	28	22	25
1963-67	20	52	40	33	23	29
1968-72	27	51	34	43	25	32
1973-77	48	64	69	62	39	30
1978-82	88	83	89	89	82	68
1983-87	81	103	90	112	85	143
1950-85 Increase	710%	45%	88%	387%	372%	921%

more inputs to boost production per acre as a substitution for land use. However, this tendency has not been evident recently. Farmers also have been under strong price pressure, with the prices received for agricultural commodities increasing at a much slower rate than the costs of the farm inputs listed above–except for the costs of pesticides and fertilizer. Farmers' resource use decisions not only reflect relative prices between inputs, but declining product prices relative to most of the inputs, as is shown in Table 3 (U.S. Department of Agriculture 1990).

Government commodity programs in the past have rewarded farmers for producing the primary commercial food and feed grain crops. Insofar as price supports tended to encourage farmers towards monocultures of these crops, the policies bear some responsibility for production patterns. However, what is critical with respect to increasing chemical use is not just monoculture, but also those aspects of government programs that limited land use and encouraged the use of more nonland inputs. At the same time, since the end of the Second World War, the shifting relative prices gave farmers strong incentives in the same direction. Pesticides and fertilizer would be increasingly used whenever possible to substitute for relatively more expensive inputs even without further incentives from government programs. This steady substitution has continued through the waxing and waning of programs.

In addition, technology has played a critical role in changing farm production, practices, and viable options. The development of fertilizer responsive varieties has greatly increased the economic value of fertilizer use and has increased use even beyond that warranted by its declining relative price. The development of continuing generations of chemical pesticides that have been successful in controlling pests has made them the effective substitute

Table 2. Quantities of farm inputs used in U.S. agriculture (indices with 1977 = 100).

	Index of		Pounds	Pounds	Index of	
Years	Crop Land	Fertilizers & Pesticides	Per Acre Nitrogen	Per Acre Phosphate	Power & Machinery	Farm Labor
1950	100	19	6	11	72	265
1955	100	26	11	13	83	220
1960	94	32	17	16	13	177
1965	89	49	31	24	80	144
1970	88	75	51	31	85	112
1975	97	83	52	27	96	106
1980	101	123	64	31	101	96
1985	98	121	67	27	81	85
1987	88	100	63	26	74	78
Increase		426%	950%	136%	3%	
Decrease	12%					71%

for other inputs–again compounding the impact of their declining cost relative to other inputs.

The invention of the tractor was a major milestone. It changed the physical management of the soil, the size and structure of farm firms, and the nature of the landscape with respect to row crops as opposed to pasture and small grains. With the tractor, no portion of the farmer's land had to be devoted to pasture, hayland, or small grains for draft animals. Farmers no longer needed the skills necessary to propagate and husband farm animals. If all farmers today had to keep some draft animals to work their farms, these skills would have been maintained and other livestock raising would also likely be more widely spread over the countryside. Tractors also were critical in allowing the expansion of the size of the farming operation, so that farming units are now sized for efficiency based on the nature and capacity of the machinery stock.

Looking back, one of the first broad investigations of U.S. agriculture to assess the state of the sector was the *Report of the Country Life Commission* (Bailey et al. 1909). This report responded to public concerns about the quality of life in rural areas. However, the remedies for the deficiencies in rural life, to a large extent, were to come from improved education, the development of modern technology for better farming, better infrastructure, and the like. The Federal Government policies that followed to promote research, technology transfer, education, and improved rural infrastructure had an important impact upon the way farmers farm today, but not through direct government involvement in markets or farming.

In contrast to these earlier concerns about the quality of life, the primary

Table 3. Farm prices received (indices with 1977 = 100).

Year	1960	1965	1970	1975	1980	1985	1960-85
Prices Index	52	54	60	101	134	128	up 146%

goal of farm programs during the Great Depression was the transfer of income to rural areas. From the beginning of the Agricultural Adjustment Act, programs for cash crops like cotton and for major food and feed grains were the best vehicle for raising income. The more environmentally benign agricultural "products" such as pasture, hayland, grassland, oats, and barley (which were closely related to draft animal use and coincident small scale livestock production) were not effective vehicles for income transfer. Therefore, the noninclusion of such products and rotations was not a failure of farm policy. Farm policy, in this instance, was not driven by a conservation goal as some believe it was or should have been (Reichelderfer 1990). While Big Hugh Bennett and other proponents of soil conservation were concerned with conservation issues and were successful in getting government policy to address them, the way farmers farmed was not the paramount issue. The conservation aspect was an instrument of convenience for other goals.

The tools utilized to raise incomes were price supports and production restrictions. The first without the second led to expenditure beyond available resources. After the initial program introduction, several critical Supreme Court decisions reshaped these mechanisms for Commodity Programs. The Hoosac-Mills decision invalidated the direct production controls provisions of the 1933 Agricultural Adjustment Act (Bowers et al. 1984). A different approach was then required and adopted.

The 1938 Act included the broadly applied nonrecourse loan concept which became the basis of price supports while applying the notion of taking land out of production for "conserving uses." This was the device that allowed some measure of production controls within the constraints set down by the court. The 1938 Act also included crop insurance provisions. During this period marketing orders were being developed, the Farm Credit system was being established, and agriculture was becoming mechanized–a process that would greatly accelerate during the Second World War. The War also left the nation with a vast unused capacity to produce nitrates once the fighting ceased. Cheap nitrogen fertilizers and tremendous incentives to develop and use fertilizer responsive varieties were a product of this war surplus.

There is no question that government farm policy helped push farmers toward today's agriculture, but so many other strong forces moved in the same direction that it is impossible to assign causality to any one policy action or influence.

CHARACTERISTICS OF FARMING AND SUSTAINABILITY

Much of the discussion of sustainability and agricultural policy reflects a desire to move from where we are now to something different. Some of the characteristics of today's farming are the following:

1. Specialized crop and livestock farming.
2. High crop prices or low input costs, encouraging particular crop choices and/or production intensification.
3. Extensive use of off-farm inputs.
4. Little necessity for concern with off-farm impacts of the production process.
5. Increasing size and concentration of production.

There are some important differences between today's farming and sustainable agriculture (Conway et al. 1991). Some are fundamental; for instance:

1. Sustainability implies less specialized farming, often requiring mixed crop/livestock farming for less dependence upon outside inputs.
2. Sustainability implies that off-farm inputs ought to be fully priced (i.e., have no direct or indirect subsidies), and that prices for crops having adverse environmental impacts should not have program price supports.
3. Sustainability implies that farm-level decision making take account of the off-farm impact of activities conducted on the farm (e.g., downstream soil deposition or water contamination costs to be borne by the farmer).
4. Sustainability (to many proponents) implies family-sized farms, not corporate factory farms.

The question arises whether changes in present farm policies can be sufficient by themselves to move farmers to practices so very different from those currently used.

FEDERAL POLICIES AND THE WAY FARMERS FARM

In a democratic society, government policies are only one piece of the action, and are constantly subject to modification and change–it matters not whether we are dealing with democratic socialism or democratic capitalism. Under capitalism, government policy towards a sector can be further constrained insofar as the free market gives demand and supply signals. However, most wealthy capitalist countries have chosen to "modify" the market for agricultural products to ensure a stable food supply, higher incomes for farmers, and (in some cases) low prices for consumers. In this sense, government influence over the sector is larger than one might first suppose in a capitalist economy. If government decides to return its agriculture to the "free market," then it forfeits the control that was previously exercised under the umbrella of "democratic agricultural policy." It is giving up an important option for influencing behavior.

If we ask what Federal policies might actually change the way farmers farm, we have to ask this with respect to some very different approaches to farm policy and also with respect to a broad array of nonfarm Federal policy as well.

Farm Policies

First, consider farm policy much along the lines we have undertaken it in the past, where the government is a major actor in price setting and supply regulation for crops. Given the voluntary participation tradition of U.S. farm policy, the program has to be attractive enough economically for a large number of farmers to join so that the impacts of the program, both the benefits and obligations of participation, can be widely distributed.

Some years ago it was believed that program provisions had great impact on farm practices. We believed that acreage restrictions greatly increased the intensity of production, including increased agrichemical use as a land substitute. Two decades ago, a 50% increase in pesticide use was projected from a 10 or 15% decrease in allowed acreage (Carlson 1990). Given current conditions and technology, we would no longer expect such an increase. One reason for this is that program prices are closer to world prices, and acreage controls are less restrictive now as compared with those periods of high price supports that occurred on a cyclical basis. Another reason is that the extremely high value in use for many agricultural chemicals has them being used already at their most effective rate. This level is often economical across a wide spectrum of costs and prices. In other cases, high chemical costs dominate decisions on levels of use. Finally, we also now have a high level of knowledge among farmers about pesticide use that lays bare the notion that more is automatically better.

What about the concern that farm programs lock farmers into growing the less environmentally benign crops? What would be the impact of a new farm program allowing 20% or, alternatively, 100% planting flexibility while maintaining income levels to farmers from previous program cropping benefits (Doering and Ervin 1990)? With 20% flexibility there is little reduction (less than 1 or 2%) in overall national agrichemical use. With 100% flexibility, nitrogen use would decline about 4% and pesticide use would fall between 2 and 3%. The potential for nitrogen leaching would drop in all regions except the Northeast, Appalachia, and the Delta. Erosion would decline in Appalachia and the Corn Belt, but increase slightly in the Delta. Nationwide soil erosion would decline, especially in the Northern Plains and the Lake States. In this analysis, a rather drastic change in the nature of the traditional program produces a remarkably modest result in terms of changes in cropping patterns and input use.

What kind of impact can we expect from an agricultural program that is "market driven"? An analysis looking at the resource-use impacts of current farm programs indicates that these price support programs result in approximately 8 or 9% increased use of capital and labor on program crops as compared with the no-program (free market) scenario. Today's farm programs do increase total chemical use by 12% on program crops–less on nonprogram crops (Shoemaker et al. 1990). However, there are still some who do ascribe much greater impacts to farm programs (WRI 1991).

The critical question is what kind of leverage can agricultural programs exert on farmers to achieve new goals not directly related to farm income, the level of commodity production, and the price of food? If program participation remains voluntary, the question is how attractive will participation be if farmers receive little price support beyond the free market and have to restrict or change input use, cropping patterns, and production practices? Within the traditional program framework, what might farm programs do to incorporate the off-farm impacts of a farmer's actions in his production decisions or to encourage more sustainable practices? These goals would require a tremendous degree of specific targeted program application through "downstream" monitoring and specific practice or activity determination. What incentives would be used to encourage such practices?

This is not to say that past farm programs have not had impact at critical times–especially in combination with other coincident forces. In the late 1970s, farm policies that subsidized inputs, stimulated production, and encouraged exports led to more intensive production at higher costs over a larger land base than would have been the case otherwise. There were clear domestic environmental and other net costs from these policies which were not mirrored in increased farm earnings, let alone in compensation to those suffering damage. Foreign food consumers were the primary beneficiaries, and some questioned whether such policies made sense (Doering et al. 1982; Schmitz et al. 1986).

Another concern is what changing the way farmers farm might do to the structure of the farm sector. One of the potential contradictions to the broad notion of sustainability is that the level of management skills and capacity for investment that frequently appears to be required for more environmentally sound agriculture is more common on large rather than small or moderately sized operations (Gadsby 1990). The structural dimension is further complicated by our current national policy not to enforce traditional anti-trust laws under the rationale that American firms are competing in broader world markets. This implies larger and more diversified processing and marketing firms beyond the farm gate with a potential for influencing the agricultural production sector towards more concentration, larger size, and more standardized products.

Nonfarm Policies

Nonfarm policies can influence the way farmers farm. Broad national policies toward the economy, trade, industry structure, resources, and the environment can have more impact on the way farmers behave than "farm" policy. Consider the following examples: low real interest rates in the 1970s encouraged overcapitalization in agriculture; the political decisions by the U.S.S.R. and China to import commodities, combined with lower value of the dollar, changed the shape and composition of U.S. agricultural exports and built an export orientation for a decade. The elimination of trade barriers under the current trade negotiations would have an impact upon the product prices received by farmers in future years and would shape the resulting intensity of resource use, choice of cropping patterns, and location of production.

National or State-level decisions about resource use, which may have nothing to do with farm policy, will determine the type of farming to be practiced in a number of locations where current systems are the antithesis of sustainable. California is the prime example of this in terms of the potential for both National and State water policy changes to affect the cost and availability of water to agriculture. National decisions about health, safety, and environmental quality–whether they be specific to agriculture or generic in nature–have had and will have great influence on the way farmers farm.

Recent Changes in Farm Policy

The flexibility provisions have been of most interest to those concerned about sustainability. However, farmers do not appear to be substantially changing their crop mix. This behavior tends to confirm the analysis that even full flexibility would have limited impact on crop mix, chemical use, and potential for water quality problems. Inertia or other forces that drive farmer decisions have not changed or may be moving in another direction. However, the modest chemical reductions projected may be more critical for environmental reasons if it is the marginal or last portion of an application that causes most of the environmental problem (i.e., that last bit that is not effectively utilized or does not effectively degrade).

The required recordkeeping for registered pesticides is a different aspect of the new legislation that may be critically important in the future. If Federal farm policies are to target specific uses and practices, such recordkeeping is essential–both as a base of knowledge about current practices and as a guide for change. The negative stance on this issue of groups like the American Farm Bureau reflect their concern about property rights and their opposition to specific regulation or control as an instrument of national policy (Kleckner 1990).

The moderate expansion of the Conservation Reserve Program (CRP) reflects the public feeling that this program meets some environmental concerns and some farm risk management and income goals. Insofar as it has targeted fragile or vulnerable areas, it has reduced environmental problems. The same can be said for the wetlands provisions. However, neither program addresses the sustainability or environmental concerns relative to those acres remaining in farming, and the long-term use of CRP acres has not been decided.

What Incentives or Disincentives Are Still the Same?

The increased flexibility under the new legislation does lessen the extent to which farmers are locked into fixed acreages of program crops to protect crop base and income payments. However, this flexibility is not as complete as that analyzed under 100% flexibility with payments maintained (Ervin et al. 1991). Thus, its impact can be expected to be even less than the modest change in farming practices projected under full flexibility.

National and international economic forces still encourage the production of commodities much as they are produced now. Other institutional arrangements already in place–research, extension education, crop insurance, credit arrangements, etc.–do not influence farmers to radically change basic practices and cropping systems. The level and character of the economic contributions by agrichemicals to production remain about the same. One analysis projects that a 20% price increase for nitrogen would result in a 5% decrease in nitrogen use, but cropping systems would remain much the same (Johnson et al. 1991). With corn, nitrogen's economic contribution is so high that it might take a several-fold increase in nitrogen prices to fundamentally change existing cropping systems. The elimination of all commodity programs and the institution of free market prices for all commodities would not significantly change these values in use characteristics for many critical inputs. A free market system by itself would also not address the problem of making the downstream impacts of farming an integrated part of the farm decision process.

What Are Our Farm Policies Designed to Do?

Writing on some essentials of a good agricultural policy in 1940, George Tolley saw three key objectives of farm policy:

1. "Activities designed to increase incomes of farmers who produce commodities for sale on a commercial scale;
2. Efforts to raise incomes and to improve the living conditions of migrant laborers, share-croppers, subsistence farmers, victims of drought or flood, and others at a disadvantage within agriculture itself; and
3. Activities designed to encourage better land use and more efficient production."

It is this last objective that included concerns within the broad area of sustainability and the environment. Tolley then commented that ". . . most government programs of both the distant and the recent past have been directed toward improvement in the condition of commercial agriculture." He hoped that the last two objectives would receive increasing attention in the future (Tolley 1940).

Today, we have the same structure of farm policies that Tolley was writing about in 1940. These instruments are still structured to best accomplish their original task, not what we now think we want Federal farm policy to do.

BRINGING ABOUT CHANGE IN THE WAY FARMERS FARM

Tinkering with conventional farm policy just will not bring about basic changes in the way farmers farm–movements at the margin, yes; basic system and structure changes, no. The institutions and program devices involved were never intended for the purposes people would like them to achieve today. Traditional policy actions will bring about only marginal responses in these areas, and if we maintain the tradition of voluntary participation, the cost of effective incentives will be high. Our historical experience with incentives has been that they affect, on the level of farm production, the relative mix of major food and feed grains, and the level and distribution of incomes to farmers. We are not dealing with the sort of situation that held in Britain for many years where the National Farmers Union would negotiate income and working conditions with the government and then commodities would flow unhindered (Self and Storing 1962). The nonfarm British public held in its own hands many decisions about environmental amenities like green belts and a host of things impacted by agriculture which are not concerns of our farm programs. As a society, we are becoming more interested in those other impacts of agriculture that have been a central concern of Britain and some other European countries for generations. These impacts are not yet part of the set of concerns addressed by our agricultural policies nor have they been a concern of the nonagricultural public until recently. What would be required in the United States would be a much more strongly held view by the general public that they should have a say in long-term decisions about land use and resources.

Under these conditions, what then is the potential or likely role of Federal policy? The critical question is whether new approaches and/or new institutions are brought to bear on the issues. In order to meet environmental goals, will expensive command and control approaches be used extensively? Realistically, the command and control approaches are almost precluded by both the level of knowledge and the intensity of administration necessary to make these approaches effective for the large number of dispersed and very different farming operations around the Nation. Will there be an effort to utilize some of the economic incentive and free market approaches like those set forth in the Clean Air Act? The question is: What new approaches and institutions can be developed that are not punitive, might be more effective than the current ones, and would bring about change?

CONCLUSIONS

Federal policies toward agriculture do not appear to play a major role as an incentive or as a disincentive for less intensive and more environmentally benign agricultural practices and cropping systems. Without any government programs or with a world free-trade situation, cropping patterns and input use will still be guided by the relative costs of inputs and resources similar to those that predominate today for the major food and feed crops. Production systems are likely to stay much the same. However, eliminating price supports will reduce intensity at the margin and will shift production regionally.

Changing the way farmers farm is not likely to be accomplished by tinkering with base acres and program flexibility. This suggests the need for developing new approaches that target specific practices, cropping patterns, input use, or sensitive locations. Assigning property rights to wetland benefits and allowing them to be sold for environmental or production uses may be an example of a targeted approach that would meet specific concerns without the costs of command and control (Lovejoy 1991). It would take only a portion of the Federal savings and loan bailout money to purchase outright all the Nation's remaining wetlands, if the public had chosen that course rather than tolerate poor regulation and mismanagement of the banking system.

The cost and complexity of the command and control approach appears unreasonable given the number of farms and the very specific resource base and management capacity associated with each one.

Federal policy regulating the availability or use of inputs on the basis of health and safety criteria does have a very real impact. However, specific farm-level application is almost impossible with this approach even if adequate knowledge could be gained about each farm's circumstances. Thus, worst-case scenarios are applied at high cost.

The alternative of taxing inputs that support unsustainable practices will vary in effectiveness depending on the value in use of the inputs. Wherever there are very high values in use, the necessary taxes and income transfers will be substantial and problematic to handle effectively.

Our task now is to invent new policy mechanisms to deal specifically and equitably with the new environmental concerns. The 50-year-old hammer should be discarded as a relatively ineffective tool, and a new one taken up that reflects society's changing values while also recognizing the actual production decisions that farmers face every day.

REFERENCES

Bailey, L.H., H. Wallace, K.L. Butterfield, W.H. Page, G. Pinchot, C.S. Barrett, and W.A. Beard. 1909. Report of the country life commission. In *Agriculture in the United States: A documentary history*, vol. 2, ed. W. Rasmussen, 1860-1906. New York: Random House, 1975.

Bowers, D., W. Rasmussen, and G. Baker. 1984. *History of agricultural price-support and adjustment programs, 1933-1985*. USDA Agricultural Information Bulletin no. 485, 11-14. Washington, D.C.: U.S. Department of Agriculture, Economic Research Service.

Carlson, G. 1990. Farm programs and pesticide demand. Paper presented at the Agricultural Economics Workshop, North Carolina State University, March 1990. 28 pp.

Conway, R., O. Doering, R. Nehring, and S. Frerichs. 1991. *A perspective on sustainable agriculture*. Draft manuscript in process, courtesy of the authors. 28 pp.

Doering, O., A. Schmitz, and J. Miranowski. 1982. *The full costs of farm exports*. Giannini Foundation Working Paper #206. University of California. 28 pp.

Doering, O., and D. Ervin. 1990. Policy options to help the environment. *Agricultural Outlook*, September: 21-22. Washington, D.C.: U.S. Department of Agriculture, Economic Research Service.

Ervin, D., and eleven co-authors. 1991. *Conservation and environmental issues in agriculture: An economic evaluation of policy options*. Washington, D.C.: U.S. Department of Agriculture, Economic Research Service, Resources and Technology Division. 62 pp.

Gadsby, D. 1990. The impacts of environmental regulations on the structure of American agriculture. Unpublished paper presented at Castelfranco, Venice, and Padova, Italy, Conference on Agriculture and Its Environment in a Normative Setting. Sponsored by Ented Suiluppo Agricolo Del Veneto Esau, May 1990, 58 pp.

Hallberg, M. 1988. *The U.S. agricultural and food system: A postwar historical perspective*, 27. The Northeast Regional Center for Rural Development, Publication #55. University Park, PA: Pennsylvania State University.

Johnson, S., J. Atwood, and L. Thompson. 1991. Tradeoffs between agricultural and chemical policies. In *Commodity and resource policies in agricultural systems*, ed. R. Just and N. Bockstael, 254-74. New York: Springer-Verlag.

Kleckner, D. 1990. Remarks by Dean Kleckner, President American Farm Bureau Federation to the American Society of Agricultural Engineers Roundtable, Chicago, IL, December 19, 1990. 15 pp.

Lovejoy, S. 1991. *Meeting America's soil and water goals without regulation*. Journal Paper #13035. West Lafayette, IN: Purdue University Agricultural Experiment Station. 8 pp.

Manchester, A. 1991. *U.S. food spending and income: Changes through the years*. Agricultural Information Bulletin #618. Washington, D.C.: U.S. Department of Agriculture, Economic Research Service. 20 pp.

Reichelderfer, K. 1990. Environmental protection and agricultural support: Are trade-offs necessary? In *Agricultural policies in a new decade*, ed. K. Allen, 203-205. Washington, D.C.: National Center for Food and Agricultural Policy, Resources for the Future.

Reichelderfer, K., and T. Phipps. 1988. *Agricultural policy and environmental quality*. Washington, D.C.: National Center for Food and Agricultural Policy, Resources for the Future. 24 pp.

Schmitz, A., D. Sigurdson, and O. Doering. 1986. Domestic farm policy and the gains from trade. *Am. J. Agric. Econ.* 68(4):820-27.

Self, P., and H. Storing. 1962. *The state and the farmer.* Winchester, MA: George Allen & Unwin. 251 pp.

Shoemaker, R., M. Anderson, and J. Hrubovcak. 1990. *U.S. farm programs and agricultural resources.* Agricultural Information Bulletin no. 614. Washington, D.C.: U.S. Department of Agriculture, Economic Research Service. 8 pp.

Tolley, G. 1940. Some essentials of a good agricultural policy. In *Farmers in a changing world: The 1940 yearbook of agriculture*, 1169. Washington, D.C.: U.S. Department of Agriculture.

U.S. Department of Agriculture. 1940. *Farmers in a changing world: The 1940 yearbook of agriculture.* Washington, D.C.: USDA. 1215 pp.

U.S. Department of Agriculture. 1985. Agricultural statistics 1985. Washington, D.C.: USDA.

U.S. Department of Agriculture. 1988. Agricultural statistics 1988. Washington, D.C.: USDA.

U.S. Department of Agriculture. 1990. *Agricultural Outlook*, Yearbook issue, April: 7. Washington, D.C.: USDA, Economic Research Service.

U.S. Department of Agriculture. 1991. *Economic indicators of the farm sector: Production and efficiency statistics, 1989*, 37. Washington, D.C.: USDA, Economic Research Service. #ECIFS9-4.

World Resources Institute. 1991. *Paying the farm bill: U.S. agricultural policy and the transition to sustainable agriculture.* Washington, D.C.: World Resources Institute. 70 pp.

Building Sustainable Agriculture: A New Application of Farming Systems Research and Extension

Cornelia Butler Flora

SUMMARY. Sustainable agriculture requires the balancing of a variety of goals. This means that often no single goal can be maximized, since such optimization might totally preclude the achievement of one of the other goals of sustainability. For this reason, transdisciplinary teams containing advocates of the various goals, with ability to negotiate priorities, provide an important input into research and extension toward a sustainable agriculture. Further, farmer membership on these teams is particularly crucial, because a sustainable agriculture means that the farmer shifts from a *user* of technology to a *producer* of technology and a *monitor* of its impacts.

A major impediment to the development of transdisciplinary approaches is the lack of good indicators of sustainability. In part this is due to the ease of use of traditional measures of production and profit. Not only do these yield single, summary measures; they have relatively short term manifestations. Indicators of the impact of agricultural practices on sustainability are more diffuse and more long term. Systematic efforts are needed to develop such measures.

Farming systems research and extension (FSR/E) has traditionally involved multidisciplinary teams, which have included farmer participation. However, both the composition and process of FSR/E must be altered to include the multidimensions of sustainable agriculture. Such adaptations are possible in all phases of FSR/E: from diagnosis, to design, to on-farm trials, to monitoring and evaluation, and finally extension.

Cornelia Butler Flora is Professor and Head, Department of Sociology, Virginia Polytechnic Institute and State University, Blacksburg, VA 24061-0137.

INTRODUCTION

Sustainable agriculture is as much a process as an end point. More than a series of techniques, it can be viewed as an approach to agriculture that attempts to find a balance among agronomic, environmental, economic, and social optimums, based on the following definition provided by Francis and Youngberg (1990):

> Sustainable agriculture is a philosophy based on human goals and on understanding the long-term impact of our activities on the environment and on other species. Use of this philosophy guides our application of prior experience and the latest scientific advances to create integrated, resource-conserving, equitable farming systems. These systems reduce environmental degradation, maintain agricultural productivity, promote economic viability in both the short and long term, and maintain stable rural communities and quality of life.

Sustainable agriculture avoids maximizing any single outcome variable, whether it be environmental quality, economic return, yield per acre, or number of family farms, in order to achieve a long term balance among the variables. Sustainable agriculture seeks a balance of environmental conservation, agricultural production, farm profit, and community well-being. Not only must research attempt to balance these multiple and sometimes conflicting goals, but it must constantly monitor and adjust the process to determine the impact of achieving one goal on the achievement of others.

The balancing act required by sustainable agriculture is made more difficult because it requires more than the usual single, summary measures of success. It will neither turn the prairie back to the buffalo nor get prize-winning corn yields per acre–although a sustainable agriculture might include large areas of extensive grazing and high corn yields. The multiple purposes of the agricultural enterprise must be acknowledged and accounted for. This can best be done when each of the goals (decreased environmental degradation, long-term yield, long-term profitability, contribution to community) is advocated and questioned by those involved in setting research agendas. A transdisciplinary approach to research, monitoring, and evaluation has the potential to achieve the desired balance. Like any living system, this balance requires constant realignment and adjustment, for the system never is truly balanced, but is simply striving to maintain balance.

One way to approach such a research agenda is to put together research and extension teams based on the farming systems and research methodology developed in the 1970s and 1980s. Farming systems research and extension (FSR/E) brings together interdisciplinary teams to diagnose the constraints to production, to design research to overcome those constraints, to carry out on-farm trials of the proposed technology, and to extend the technology once it has been proven effective in increasing productivity on farmers' fields under farmers' conditions. This methodology can be modified to develop technology to overcome constraints to sustainability on the farm level. It also can be used to identify other constraints to the implementation of a more sustainable agriculture: the absence of marketing systems, appropriate credit, and pricing systems, or agricultural and macro policies that influence the acceptance or rejection of a new technology at the farm level.

This article examines how the implementation of FSR/E would take place when oriented toward sustainability. The composition of the teams, the manner in which they would function, and the potential benefits are described, and some of the difficulties facing such teams are discussed. These difficulties include multiple dimensional indicators of success as well as the difficulties of transdisciplinary communications and cultural gaps.

SUSTAINABLE FARMING SYSTEMS RESEARCH AND EXTENSION

Research toward a more sustainable agriculture requires a systems perspective in order to be sure that no single goal is stressed to the detriment of the others. Thus, experts must be included in the research development team who advocate macro-level goals such as environmental conservation and enhancement, high productivity, maintenance of farm families on the land, and the development of viable communities. Research to build a more sustainable agriculture also requires high farmer participation because the farm family embodies the complexity of multiple goals on the micro level. However, when the cropping system involves hired farm labor, they must be included in identification of constraints to sustainability. While it may be possible to have an environmentally sound agriculture with only a few large corporate farms, many who are advocates of sustainable farming insist that the option of family farming must also be sustained and enhanced. It is assumed that the product of sustainable agricultural research will be a series of technologies, from cultural practices to bedding materials to new varieties, that are implemented on the farm level. It is also assumed that each technology that could be implemented to make agriculture more sustainable on a given farm or field has a series of on-farm and off-farm constraints to its adoption. Those constraints may be other technologies already in use on the farm, a lack of marketing opportunities, the lack of complementary technology, farm policies (such as those that force one to monocrop in order to maintain the security of one's base acres), or macroeconomic policies such as monetary policy aimed to cheapen exports (Flora 1990).

A sound research agenda must include a representative cross section of producers. Sustainable agriculture would have a very different configuration if the producer members of the team were limited to executive farm managers for large corporations rather than if they consisted solely of women and men who run farms they themselves own or rent. Whatever one's definition of sustainable agriculture, if the technology to make it achievable is to be adopted, we have to take into account the people involved. From their various positions in social systems, people should negotiate to determine research priorities, participate in the research, make the technology generated by that research available through both the private and public sector, choose whether or not to implement it, and consume the product. The welfare of the user, both the farmer and farm worker who implement the technology and the indirect user who buys the agricultural product or drinks the water from the watershed, must be part of our *ex ante* calculations when we invest scarce resources in research.

Problems of Measurement

Part of the negotiation process to achieve the balance that yields a sustainable agriculture is to agree on indicators of sustainability. How can we measure the degree to which one practice is more sustainable than another? And how can we make sure that multiple measures of sustainability are included when we evaluate the utility of new technology, especially when some of the impacts take more time to be discerned than others? Traditionally, agricultural experiments have focused on various measures of productivity which involve output per unit of input: grain per acre, weight gain per calf, or percent of seeds that germinate. Although sometimes difficult to systematically count, the indicators of productivity provide concrete ratio data that represent phenomena generally recognized as beneficial. They are easy to understand and to calculate. Randomized block experimental design, combined with parametric statistical techniques, could reassure the researcher that practice "*y*" yields more than an identical plot without "*y*."

But even those reassuring calculations do not reveal whether a practice is more profitable, or, an even more complex determination, how much of an input results in how much increase in profitability. Therefore, agronomic experimental designs had to be adjusted to include profitability. The logic of including economists on farming systems teams and in the design of conventional field trials is easy to grasp. Including other disciplines is more difficult, particularly those that bring in alternative measures of experimental success.

Profitability is a relatively clear concept since it reduces all inputs and outputs to a single common indicator: dollars–a generally agreed upon good. Costs are in terms of what production costs the producer directly, although there are still problems in assessing the correct value for land and labor. Traditional farm management calculations do not include such "externalities" as water quality, soil loss, or decline in genetic diversity. These collective costs are ignored, not necessarily because they are felt to be unimportant, but because of the lack of a single summary measure. What is the value in dollars of an inch of top soil? How does one systematically measure it in experiment station plots or on farmers' fields? There is no market for soil per se, and studies have shown that the impact of soil loss on the value of the farm is generally marginal (Buttel and Swanson 1985).

Sustainability has no single, summary indicator. Would it be appropriate to measure sustainability by water quality (presence of pesticides or nitrogen in surface or ground water at different times in the production cycle)? By amount of soil loss? By the increase in soil microorganisms? By use of nonrenewable resources per hectare? By change in number of earth worms per cubic meter of soil? Yet these are only indicators of environmental impacts. Impacts on family well-being (beyond short-term profitability) are difficult to measure, and impacts on community well-being even more so. What kind of social indicators can be accepted by agricultural researchers, and how will experimental methods have to be changed in order to include these indicators as dependent variables?

Further, traditional statistical tools are not adaptable to complex systems. The statistical methods traditionally used in experimental agricultural research are biased toward testing different versions of monocultures and are unable to cope with more complex system changes. Research on sustainable agriculture should include research in statistics, particularly multivariate statistics. In addition, research on computer software is necessary to empower farmer-researchers to easily analyze the significance of changes in the measures of sustainability they choose to examine.

Farmer/Researcher Partnership

FSR/E is an approach to agricultural research that attempts to include human and systemic factors (Amir and Knipscheer 1989; Hildebrand 1986; Shaner et al. 1982). FSR/E techniques can improve sustainable agriculture research (Ikerd 1990; Butler and Waud 1990).

A farming systems approach might be called a "marketing approach."

Researchers find out what the demand is and then develop the technology, rather than developing the technology and depending on the sales staff–Extension–to create the demand by convincing people they have a problem that this product just happens to solve. A farming systems approach based on sustainability would help in identifying new problems that have not been articulated because the possibility of a solution has not been present.

One immediate barrier to using a demand approach involving farmers in sustainable FSR/E is that, initially, few farmers want sustainable technology. Most sustainability problems are experienced collectively, rather than individually. Farmers, like the rest of us, are prone to identify problems based on the solutions they know are available. Further, we tend to confuse problem with solution. Thus, we might define the problem as "low corn yield," which suggests increasing inputs of nitrogen and pesticides, rather than as "low return on this field," which might suggest looking at an alternative crop or rotation. Or we might identify "lack of credit to purchase fertilizer" as a problem when fertilizer is just one of several solutions to the problem of providing nitrogen to a field or of increasing the net profitability of that field. Therefore, we have to ask the questions in ways that do not *a priori* define the answers.

A second barrier is that although FSR/E has been quick to adopt the slogan of sustainable agriculture, those involved in FSR/E practice have yet to systematically analyze how including sustainable agriculture would change the way farming systems research and extension is usually done. Some have assumed that since FSR/E is systemic and presumably involves the farmer, it will magically produce more sustainable systems. However, farmers involved in developing system level solutions can put narrow limits on their systems and impose a short time span for achieving only agronomic or economic goals. There is clear evidence that a great deal of FSR/E has focused on constraints to production and has been aimed primarily at increasing the use of purchased inputs to overcome constraints.

However, the tools of farming systems research, with their inclusion of multidisciplinary teams and their focus on farm-level impacts and farmer participation, can be redesigned to generate research and extension oriented toward sustainable agricultural systems. FSR/E has conventionally been practiced as a four-stage process: diagnosis, design, on-farm trials, and extension (Gilbert et al. 1980; Shaner et al. 1982). Let us briefly consider how each stage would be implemented to involve farmers in the research and extension process of developing technology to make agriculture more sustainable.

Diagnosis

The current practice of diagnosis in FSR/E involves interdisciplinary teams visiting the farm, generally focusing on a particular field, then determining the constraints to production on that field. (The goal is assumed to be increased production, although often followup studies of why a particular technology was not adopted suggests that maximizing production is not always the goal of the farm family for a particular field.) This technique of problem diagnosis is often referred to as *sondeo*, or rapid rural reconnaissance.

Diagnosis based on sustainability has different goals than diagnosis based on productivity. Sustainability is relatively abstract and long term. Productivity is relatively concrete and short term. A challenge in the diagnosis is to get farmers and researchers to agree that *x* would be more sustainable than *y*, and then work together to determine the constraints to implementing *x*. Additionally, there is the constant realization that a practice that may be more sustainable in terms of the environment may be less sustainable in terms of economic returns or in terms of family well-being unless policies are adopted to mitigate that contradiction. Teams must continually negotiate goals as well as means in determining a research and action agenda for sustainability.

A sondeo based on sustainability as a first goal would be somewhat different than the FSR/E tradition (Hildebrand 1981). Instead of looking first at constraints to productivity or even profitability, the analysis would identify the constraints to sustainability on a given farm. A diagnostic team would include the farmer or even the entire farm family as members. It could also include an ecologist, soil scientist, animal scientist, sociologist or anthropologist, and agricultural economist. Each would identify major constraints to sustainability from their disciplinary perspective, then negotiate the collective perspective based on achieving a balance of environmental, economic, and social optimums.

Conventional farming systems teams might talk to a farmer and determine that a major perceived problem is low corn yield. The team then would visit a stand of corn and determine, through looking at the leaves and other visual indicators, that the problem is lack of nitrogen. They might analyze the soil to confirm a low level of available nitrogen. Because it is a farming systems approach, the research team attempts to overcome the constraint within the structure of the farm and the socioeconomic status of the farmer. The team might recommend application of nitrogen fertilizer if the farmer has a relatively high cash income, the application of animal manure if the farmer runs a mixed crop and livestock operation, or the planting of a legume if the farmer has neither cash nor animals. Thus the goal of increased productivity is served through removal of the constraint of available nitrogen.

A team that included sustainability as its goal would also address such issues as soil and water conservation, alternative labor demands of different cropping systems and patterns, and the community implications of different input mixes.

Design

The farmer might point out that a constraint to decreasing use of nitrogen fertilizer is accompanying decrease in corn yield. Nutrient status would be the constraint to sustainability. But here instead of simply recommending to the farmer the amount of nitrogen fertilizer that should be applied, the farming systems team would work with the farmer to come up with alternatives to providing the required nitrogen. Will a crop rotation with legumes provide the available nitrogen? A farmer might then experiment on small plots and determine if it would. The next constraint to sustainability is the fact that if the rotation is put into place, the farmer will lose her base acres in corn and thus her options to participate in government farm programs. Thus a policy constraint to sustainability has been identified. And it is possible that the diagnosis phase would suggest that there is no way to get high corn yields without substantial use of nitrogen fertilizer. Alternative crops or even the introduction of pasture with livestock enterprises could be considered, with an *ex ante* analysis of the alternative costs and benefits involved. Often the constraint to alternative enterprises is the absence of a viable market. At that point, larger institutional issues would have to be addressed. Transdisciplinary teams oriented toward sustainable agriculture would evaluate production practices in terms of the current economic and policy environment, understanding that these may vary in the future. If the goal were simply productivity, the other alternatives and the links with policy would not have to be analyzed.

On-Farm Trials

Once a diagnosis based on constraints to sustainability is complete, on-farm trials would be designed very much as they are in conventional FSR/E. The farmer and the researchers determine if any on-the-shelf technology exists to overcome the constraint. If there is no on-the-shelf technology available to increase availability of nitrogen to corn, researchers might be called in to consider developing alternative technologies: using genetic engineering to facilitate biological nitrogen fixation in corn, using an understanding of soil physiology and microorganisms to develop improved methods for capturing atmospheric nitrogen, or other creative ways of linking concrete farmer-identified constraints to new scientific enterprises which have as their ultimate goal increased environmental quality.

There needs to be a close tie between the on-farm trials and the more basic experiment station research. The experiment station researcher comes to the farmers to discover their constraints to sustainability and how the experiment station's research can be helpful in overcoming these constraints. One of the major contributions of FSR/E is that experiment station research is more responsive to farmers' needs. With the implementation of sustainability-oriented FSR/E, experiment station research would become more responsive to the multiple needs of the environment and community as well as the farm family.

A final aspect of on-farm testing of a sustainable farming system is the need to develop a series of non-traditional measures to attempt to evaluate environmental impact as well as production costs, yields, profitability of the entire system, etc. For example, a series of simple soil traps could be put into each field in order to observe the amount of soil erosion. Measures of percolation of nutrients and chemicals to ground water and surface water could be recorded as could other measures of changes in environmental quality that can be related to particular agronomic or animal husbandry practices.

Extension

As with conventional FSR/E, once adequate technologies have been identified, tested on the experiment station, and tested on farm fields, they would be extended to *recommendation domains*. A recommendation domain is "a group of farmers with roughly similar practices and circumstances for whom a given recommendation will be broadly appropriate" (Byerlee et al. 1982). The recommendation domains, which determine which farmers would be offered which alternative technologies, would be based on the agro-ecological conditions and on the characteristics of the farmers involved, such as part-time, full-time, old, young, and availability of labor. The farmers would then adapt the technology through farmer-managed demonstrations.

Sustainable FSR/E, even more than productivity-driven FSR/E, must be viewed as an ongoing process of farmer as researcher, as the farmer constantly tests new technologies and new cropping mixes on a whole-farm basis in order to fine-tune the broader, less specific information that comes from the Extension Service. If Extension were to adopt a sustainability orientation instead of a productivity orientation, much more time would be devoted to teaching farmers how to run experiments and to identifying successful technologies being used by farmers that could be tested by other farmers. Less time would be devoted to hierarchically handing down technology established by the experiment station.

A sustainable FSR/E process would also be different from the current system of the Soil Conservation Service, which has identified a series of best management practices that everyone is supposed to adopt. With a sustainable FSR/E process, there would be much more concern for adapting those generally identified practices to the specific situation of farmers. In addition, the practices would be constantly evaluated in terms of their cost effectiveness and their environmental impact.

Sustainable FSR/E, as with the more conventional farming systems approach, would focus on interactions within the farming system. Extension would have to be based on the premise that one is moving toward sustainability, not that by changing a few practices one is suddenly sustainable. A change in one part of the system implies that all other parts adjust to it. For example, a change in the type of manure pit has implications for the way that animals are bedded in the barn and the way manure is handled and spread on the field. This wide variety of systems implications requires constant evaluations of the impact of changing specific practices on overall sustainability as well as profitability.

As another example of interrelationships in design, some of the supplements given to cattle to increase their rate of gain contain trace minerals. When these are passed through to the manure pile and spread onto the fields, they tend to accumulate toward toxic levels, according to ongoing work by the North Carolina Department of Agriculture. Thus, if one is going to start using manure on the fields, one also may have to change the diet and supplements given to one's cattle in order to maintain the sustainability of the entire system. Attention to a sustainable farming system would include an evaluation of the whole content of what is put on fields. Research that begins as a soil fertility question becomes one of animal nutrition.

Conducting sustainable FSR/E will be different from conducting the more conventional FSR/E. Diagnosis will be different, measurements for on-farm trials will be different, the attention to systems linkages will be different, and extension will be different as farmers are increasingly empowered to analyze their own systems and to conduct experiments under their own conditions. These farm-specific experiments/demonstrations may not all have the strict controls necessary for scholarly publication, but will serve to inform and further educate farmers as they continually seek to decrease both the cost of production and the negative impacts on the environment, while still maintaining a profitable farming operation that yields good water, good soil, and good wildlife habitat.

Choosing a Research Topic

Clearly, all research will not be carried out on farmers' fields. Some research vital for developing a more sustainable agriculture will involve basic scientific research to determine exactly how a process works. But basic research should be motivated by a desire to illuminate a particular constraint to sustainability. It is important that those doing the basic work in the laboratory visit farmers' fields and discuss with them their constraints to sustainability, so that researchers can make more direct links between basic science and its societal implications. Transdisciplinary inputs are necessary at this stage as well. The presence of a variety of disciplines on the teams that evaluate research proposals and award competitive grants is crucial in maintaining the "balancing act" necessary to develop technology which is appropriate for a sustainable agriculture, as well as to monitor the impact of the adoption of different agricultural technologies in different settings.

Scientists usually have many choices of research problems. Research by Busch and Lacy (1983) suggests that much of current problem choice is guided by potential for publication and for grants. Reward systems need to be reexamined to provide motivation to pick research topics with direct implications for sustainability. Transdisciplinary teams need to provide inputs into reward decisions.

CONCLUSIONS

The research and extension system, including both public and private aspects of technology development and diffusion, has a number of built-in biases against systematically addressing agricultural sustainability. Transdisciplinary teams that include farmers and other agricultural practitioners have the potential of keeping the multiple goals of sustainability in focus.

Such teams are necessary in order to begin to develop systematic measures of the various types of sustainability that can be used to evaluate the impact of different technologies. There is overwhelming pressure to evaluate changes in farming systems using traditional measures, which are biased toward economic and productivity goals. While sustainable agriculture cannot ignore these ingredients of sustainability, it is easy for these indicators to overwhelm the more fragile measures with less consensus behind them.

Research that hopes to improve the quality of the environment and the quality of rural life is by its nature long-term. The time horizon for funding research and monitoring its impact must be long enough to ensure that changes toward a more sustainable agriculture are taking place and that all the factors that go into sustainability are addressed. Transdisciplinary efforts can contribute to that process through providing collective memory and continuity.

Movement toward a more sustainable agriculture cannot just be more of the same. The kind of research that involves farmers in determining which practices are more sustainable on their farms will require different mechanisms of evaluation, resulting in fewer of the publications that ensure researchers continuing tenure in their research institutions. Further, multidisciplinary publications are rarely valued in discipline-based decisions about promotion and tenure. Universities must reevaluate how they set their research agendas and their methods of measuring research and extension productivity. The research, teaching, and extension systems in our society must become more transdisciplinary as well.

FSR/E presents a systematic methodology for changing farming systems, including changing the actions of farmers, researchers, and extension personnel. It has a tradition of involving interdisciplinary teams in its implementation. However, that methodology has derived from production agriculture. Key changes in team formulation and dynamics would be needed for FSR/E to function as a major mechanism for generating and implementing more sustainable agricultural systems. Broadening the composition and mandate of the transdisciplinary teams would be one mechanism for such change.

REFERENCES

Amir, P., and H. C. Knipscheer. 1989. *Conducting on-farm animal research: Procedures and economic analysis.* Petit Jean Mountain, AR: Winrock International Institute for Agricultural Development and International Development Research Centre.

Busch, L., and W. B. Lacy. 1983. *Science, agriculture, and the politics of research.* Boulder, CO: Westview Press.

Butler, L. M., and J. Waud. 1990. Strengthening extension through the concepts of farming systems research and extension (FSRE) and sustainability. *J. Farming Systems Research and Extension* 1(1):77-92.

Buttel, F. H., and L. E. Swanson. 1985. Soil and water conservation: A farm and public policy context. In *Conserving soil: Insights from socioeconomic research,* ed. S. Lovejoy and T. Napier, 26-39. Ankeny, IA: Soil Conservation Society of America.

Byerlee, D., L. Harrington, and D. L. Winkelmann. 1982. Farming systems research: Issues in research strategy and technology design. *American Journal of Agricultural Economics* 64(5):897-904.

Flora, C. B. 1990. Policy issues and agricultural sustainability. In *Sustainable agriculture in temperate zones*, ed. C. A. Francis, C. B. Flora, and L. D. King, 361-80. New York: John Wiley & Sons.

Francis, C. A., and G. Youngberg. 1990. Sustainable agriculture: An overview. In *Sustainable agriculture in temperate zones*, ed. C. A. Francis, C. B. Flora, and L. D. King, 1-23. New York: John Wiley & Sons.

Gilbert, E. H., D. W. Norman, and F. E. Winch. 1980. *Farming systems research: A critical appraisal.* Rural Development Paper, no. 6. East Lansing, MI: Department of Agricultural Economics, Michigan State University.

Hildebrand, P. E. 1981. Combining disciplines in rapid appraisal: The Sondeo approach. *Agric. Admin.* 8:423-32.

Hildebrand, P. E., ed. 1986. *Perspectives on farming systems research and extension.* Boulder, CO: Westview Press.

Ikerd, J. E. 1990. A farm decision support system for sustainable farming systems. *J. Farming Systems Research and Extension* 1(1):99-108.

Shaner, W. W., P. F. Philipp, and W. R. Schmehl. 1982. *Farming systems research and development: Guidelines for developing countries.* Boulder, CO: Westview Press.

Ecological Sustainability in Agricultural Systems: Definition and Measurement

Deborah Neher

SUMMARY. "Sustainable agriculture" has emerged as the most agreed-upon term to synthesize a variety of concepts and perspectives associated with agricultural practices that differ from those of conventional production. Definitions of sustainable agriculture contain three equally important components: environmental quality and ecological soundness, plant and animal productivity, and socioeconomic viability. The Agroecosystem component of the Environmental Monitoring and Assessment Program is developing a systems-level approach to the long-term monitoring of agroecosystem sustainability. Measurements will be made for a suite of indicators at sites selected from a probability sampling frame. Associations between indicator values over time will be used to assess agroecosystem condition and status on a regional and/or national scale. One or more measures of sustainability will be developed by organizing indicators and assessment endpoints into a framework based upon the three components of sustainable agriculture.

INTRODUCTION

Conventional agriculture in the United States includes capital-intensive monocultures; continuous cropping; a substantial reliance on manufactured

---

Deborah Neher is a Research Associate at North Carolina State University, 1509 Varsity Drive, Raleigh, NC 27606.

This work was supported by funds from EMAP-Agroecosystems, a cooperative program between the U.S. Department of Agriculture (USDA)-Agricultural Research Service, the U.S. Environmental Protection Agency (EPA), and a Specific Cooperative Agreement between the USDA-ARS and the North Carolina Agricultural Research Service. Although this work was funded in part by the U.S. EPA, it has not been reviewed formally by the U.S. EPA peer and policy review and may not reflect the views of the Agency.

The author thanks Walter W. Heck and C. Lee Campbell for reviewing the manuscript.

inputs such as fertilizer, pesticides, and machinery; as well as an extensive dependence on credit and government subsidies. Perceived as businesses, farms are often operated with a priority of maximizing short-term profits. Although not quantified, the ecological cost of developing and maintaining U.S. agricultural systems has been high. For example, one-third of the top-soil on U.S. agricultural land has been lost over the last 200 years (Edwards 1990). In addition, soils have become compacted and lost fertility; ground-water has been depleted and polluted from pesticides and fertilizers; wildlife habitats have been lost or damaged due to chemical runoff; and forests, range, and wetlands have been converted to croplands. Frequent use of some pesticides has resulted in the development of resistant strains of pests and pathogens, which has led to the need for more or different pesticides and has increased costs (Edwards 1990). Most of these high-input systems, sooner or later, will probably fail because they are neither economically nor environmentally sustainable over the long term (Parr et al. 1990).

Interest in sustainable agriculture has increased during the last decade. This interest has been fostered by increasing consumer concern for food free of pesticide residues, farmers' concern for their own health and that of others, and the concern of the public and policymakers about the degradation of the natural environment through various conventional agricultural practices in the United States. The term "sustainable agriculture" has been used in many different ways both in scientific papers and in popular news, and its meaning has become obscure (Lockeretz 1988). This result may be partly due to the multitude of components that comprise sustainable agriculture. The purpose of this paper is to identify common themes in the many definitions of sustainable agriculture, place the themes in an ecological context, and discuss measurable indicators and assessment endpoints that have potential for use in monitoring the sustainability of agroecosystems.

DEFINITION OF SUSTAINABLE AGRICULTURE

Evolution of the Term

The science and practice of sustainable agriculture is as old as the origins of agriculture, although the contemporary use of the term evolved more recently (Altieri 1987). Pioneers of "sustainable agriculture" were Franklin King, Lord Northbourne, and Lady Eve Balfour. In 1911, King published *Farmers of Forty Centuries: Permanent Agriculture in China, Korea and Japan*. His book documents how farmers in parts of East Asia worked fields for 4,000 years without depleting the fertility of their soil (Reganold et al.

1990). He compared the low-input and sustainable approach of oriental agriculture with what he perceived as the reckless and wasteful methods used by U.S. farmers (Stenholm and Waggoner 1990). King conveyed the idea that agriculture could not be sustained over the long-term in economic, biological, or cultural terms unless it was "rooted firmly in frugality and recycling of fertilizer elements and organic materials" (Stenholm and Waggoner 1990). Lord Northbourne was the first to use the term "organic farming" in his book *Look to the Land*, published in 1940. His vision of the farm was "a sustainable, ecologically stable, self-containing unit, biologically complete and balanced–a dynamic living organic whole" (Scofield 1986). The phrase "sustainable agriculture" was not used until the late 1970s, when it was coined by Lady Eve Balfour (Rodale 1990). Dick and Sharon Thompson (Boone, Iowa) began conducting on-farm research on "organic farming" in the 1960s, and research centers, including Rodale Research Center (Emmaus, PA), The Land Institute (Salina, KS) and Washington University (St. Louis, MO), emerged in the 1970s (Bidwell 1986). The terms "lower input agriculture" and "low-input/sustainable agriculture" (LISA) were coined in the 1980s by Clive Edwards and Dennis Oldenstadt, respectively (Madden 1989).

"Sustainable agriculture" has emerged as the most agreed-upon term to synthesize a variety of concepts and perspectives associated with agricultural practices that differ from those associated with conventional production agriculture. "Low input" or resource-efficient agriculture focuses on the resource dynamics of the agroecosystem. Other perspectives emphasize the social and ecological aspects (e.g., agroecology) (Altieri 1987), a specific set of practices (e.g., organic farming) (Lockeretz 1988), or management concepts combined with an ecological/social overview (e.g., biodynamics and permaculture) (Hauptli et al. 1990).

The definition of sustainable agriculture adopted by the American Society of Agronomy is "one that, over the long-term, enhances environmental quality and the resource base on which agriculture depends, provides for basic human food and fiber needs, is economically viable, and enhances the quality of life for farmers and society as a whole" (Schaller 1990). This definition is similar to one proposed by Altieri (1987) in which "sustainability refers to the ability of an agroecosystem to maintain production through time, in the face of long-term ecological constraints and socioeconomic pressures." Both of these definitions correspond to the relationship of agriculture to indigenous cultures; agriculture was developed to "even out environmental and economic risk and maintain the productive base of agriculture over time" (Altieri 1987).

Three common themes occur in definitions of sustainable agriculture:

plant and animal productivity, environmental quality and ecological soundness, and socioeconomic viability. All three aspects must coincide before sustainable agriculture is possible. A system must be ecologically sustainable or it cannot persist over the long run, and thus cannot be productive and profitable. Likewise, a system must be productive and profitable over the long run, or it cannot be sustained economically (Altieri 1987; Ikerd 1990)–no matter how ecologically sound it is (Stenholm and Waggoner 1990).

Method, Myth or Philosophy?

Sustainable agriculture is an approach or a philosophy (Luna and House 1990; Schaller 1990) that integrates land stewardship with agriculture. Land stewardship is the philosophy that land is managed with respect for use by future generations. Since factors that determine sustainable agriculture are measured on a time scale of decades or generations, future generations may be best able to evaluate whether or not their predecessors practiced sustainable agriculture. However, a predictive mechanism for determining sustainability of agroecosystems would be helpful *now*.

A perplexing attribute of sustainable agriculture is that there is no precise, set formula that applies to all situations. Sustainable agriculture is not simply a list of methods (Luna and House 1990; Schaller 1990) or crop production with reduced use of agricultural chemicals (Stenholm and Waggoner 1990). Sustainable agriculture is applied uniquely to each site and is a management-intensive, resource-conserving process that considers both long- and short-term economics (Stenholm and Waggoner 1990). Although sustainable agricultural practices must be tailored to specific regions, soil types, topography, and climate (Lockeretz 1988), ten general attributes (1-9 from Lockeretz 1988; 10 from Hauptli et al. 1990) may be associated with the concept of sustainable agriculture. These include: (1) crop varieties and livestock selected for suitability to a farm's soil and climate, as well as for resistance to pests and pathogens; (2) livestock housed and grazed at low densities; (3) farm-generated resources preferred over purchased materials since the former are generally renewable; locally available off-farm inputs preferred over those from distant regions because the former require less energy in transport; (4) diversity of crop species desired for stability and achieved by rotations, intercropping, and relay cropping practices; (5) rotation of crops, enhancing utilization of nutrient reserves in lower soil strata by including deep-rooted crops, and aiding in control of weeds and pests; (6) cover crops and mulching used to reduce erosion and to conserve moisture by protecting the soil surface; (7) soils managed to increase their ability to hold nutrients and to release them at an appropriate time for crop utilization; (8) soluble inorganic fertilizers applied at a level that a crop can use efficiently, and only to the extent that nutrient deficits cannot be met by livestock manures and legumes; (9) synthetic pesticides used to enhance control of weeds, insect pests, and pathogens, but only as a last resort when there is a clear threat to the crop; and (10) biocides, when employed, targeted to specific organisms and meeting the criteria of low mammalian toxicity, limited persistence, and low environmental mobility.

Sustainable agriculture requires increased knowledge about and management of ecological processes. In conventional U.S. agricultural practices, ecological processes viewed as necessary for sustainability may be disrupted or altered by large inputs of agricultural chemicals. For example, use of insecticides may increase weed populations by removing natural enemies of weeds; application of fungicides may act on nontarget soil fungi that provide a natural control of nematode population levels; and use of insecticides and fungicides may reduce earthworm populations, thus lowering soil fertility and water infiltration rates (Luna and House 1990).

Conservation of resources is essential to permit long-term use of agricultural lands. Conservation of resources should not be confused with *preservation* of resources. Conservation implies the wise "use" of resources and assumes an understanding of the difference between renewable and nonrenewable resources (Schaller 1990). Ecological resources must be recycled–not depleted–or agriculture cannot persist.

MONITORING AGROECOSYSTEM SUSTAINABILITY

Ecosystem Perspective

Agricultural systems are "ecosystems." By definition, an ecosystem is a unit composed of associated communities of organisms and their physical/chemical environment. By intention, people represent one of the communities in agroecosystems and are *not* external to ecosystem functions. People play a governing role in regulating agroecosystem processes, some that lead toward and others that impede sustainability in agroecosystems. People are responsible for the selection of crop varieties and livestock breeds, and they impart techniques, social organizations, values, and knowledge to the function of agroecosystems. Ecosystem-level concepts require "systems-level" thinking and research; systems are inherently complex, with a multitude of interactions. A "systems" approach to studying and measuring agroecosystem structure and function is interdisciplinary and includes biological, chemical, physical, and social scientists. Ecosystem-level concepts are the *core* of sustainable agriculture–both in definition and measurement!

A major challenge in monitoring agroecosystem sustainability is to have indicators that identify system function or dysfunction at scales ranging below and above individual ecosystems (i.e., populations to global systems) (Gliessman 1990). At all scales, the three components of sustainable agriculture (environmental quality, plant and animal productivity, socioeconomic viability) are confounded with at least four important ecological processes: nutrient cycling, hydrology, population dynamics, and energy flow. Below are some examples of quantifiable attributes of the three components of sustainable agriculture in relationship to ecosystem structure and function.

Environmental quality and ecological soundness in agroecosystems can be monitored by measuring selected indicators of nutrient cycling, hydrology, and resource conservation. Persistence of life requires recycling of nutrients between living organisms and the physical environment. Several measures are used to assess and monitor cycling of nitrogen within agroecosystems. Measures of nitrogen inputs to an agroecosystem include the amount of nitrogen in rainfall, and the timing and rate of application of chemical fertilizers, animal manure, or sewage sludge. Two factors that affect the conversion of nitrogen to useable forms by plants include (1) the use of legumes in crop rotations and (2) populations of microfauna that graze on microbial decomposers (i.e., bacterial-feeding nematodes) (Freckman 1988). In agroecosystems, nutrients are removed from their cycles as harvested products (King 1990) or are exported from the field through leaching, denitrification, volatilization and erosion. Nutrients must be replaced in order for agricultural production to continue at an economically viable level. Measures to determine nitrogen removal include erosion, depletion of organic matter, and chemical exports from the field.

Availability of water (without toxic levels of contaminants) at appropriate times and locations is essential to the sustainability of an agroecosystem. Of course, water is essential for the physiological functioning of biological organisms. Water flow affects nutrient inputs and losses through leaching and erosion. Factors that regulate water cycling include inputs such as precipitation, run-on and irrigation, and outputs such as runoff, infiltration, and mechanical drainage (e.g., subsurface tiles). These factors can be measured, as can the quality of ground and surface water (e.g., salinity, presence of agricultural chemicals, and other contaminants).

Ecological soundness includes maintaining the physical, chemical, and biological integrity of soils. In sustainable agriculture, it is important to maintain a certain level of crop production along with diversity and well-being of soil-inhabiting organisms (Hauptli et al. 1990). Ecological integrity of soils can be achieved by balancing degradative processes such as soil erosion, nutrient runoff, and organic matter depletion with the beneficial effects of crop rotation, conservation tillage, and recycling of animal manures and crop residues. The challenge is to achieve a balance between the degradative and beneficial processes (Parr et al. 1990) so that the three basic attributes of sustainability are realized. A useful index of soil quality should integrate physical, chemical, and biological parameters that quantify the relationship between degradative and beneficial processes.

Nematode community patterns could provide an indication of the biological health of soils. Omnivorous and predaceous nematodes provide "connectedness" to the detritus foodweb (Coleman et al. 1983) and lengthen food-chains, respectively; their presence and/or abundance reflect agroecosystem stability (Wasilewska 1979). Bactivorous nematodes are important regulators of decomposition because they feed on microbial decomposers (Freckman 1988). Abundant populations of plant-parasitic nematodes may limit crop productivity by consuming primary production (Wasilewska 1979).

Population dynamics of organisms are important in maintaining a sustainable balance between populations, their respective food sources, and space requirements. Population dynamics of crops are regulated by factors including hydrology, soil structure and fertility, climate, fertilizers, crop diversity, and other organisms. Population dynamics of insects, pathogens, and weeds are regulated by predator-prey interactions, competition, mutualism, and human activities including cultural, chemical and biological control practices. Toxic agricultural chemicals and metal contaminants may influence human population dynamics. Measures that could be used to monitor regulation of biological populations include applications of herbicides, pesticides, and fertilizers; populations of insects (both beneficial insects and pests); employment of cultural control strategies, including selection of crop varieties and livestock breeds; and management practices that influence biological diversity.

Biological complexity and diversity, which are essential components of sustainable agriculture, require the maintenance of a wide range of plant types and habitats on the farm (Hauptli et al. 1990). Increased diversity of species within an agroecosystem (e.g., polycultures, hedgerows) may decrease risks of production failure by providing alternate crops and by promoting natural predators of pests (Hendrix et al. 1990). The establishment and maintenance of complexity and diversity require a sophisticated understanding of population dynamics to manipulate relationships among hosts, pests, and predators in the agroecosystem. This manipulation, in turn, serves to minimize major disruptions that require other kinds of intervention (e.g., pesticide applications for pest management) (Hauptli et al. 1990). Indicators of biodiversity include (1) employment of management strategies such as strip-cropping, crop rotation, trap crops, inter-cropping and multilines; (2) indices of diversity and fragmentation of agricultural landscapes; and (3) the quality of wildlife habitats.

Productivity is a measure of energy flow, the foundation of an ecosystem. A productivity index can be developed to reflect the energy efficiency of production by differentiating renewable and nonrenewable sources of energy. This requires conversion of all ecologically important inputs and outputs to a common currency. Ecologically important inputs include solar radiation, human labor, work of machines, fertilizers and herbicides, seed, hay, irrigation water, pollutants, and pesticides. Ecologically important outputs include plant and animal products (grain, vegetables, meat, milk, etc.), chemical exports, and sediment loss (Olson and Breckenridge 1992). A variety of currencies exist for comparing inputs and outputs. A common currency may take the form of net primary productivity, net caloric output per caloric input, protein output per unit caloric input, or standardized dollar values.

Socioeconomic indicators may reflect the quality of life and profitability for farmers, farm workers, and rural communities. Farm-level indicators include the operator's age, farm size, whether a farm is managed by the owner or a hired manager, and level of indebtedness. Field-level indicators include land tenure, fuel costs in cultivation, pesticides and fertilizers, and person-days of hired, custom, or family labor. The challenge is to select indicators that reflect the well-being of people and the environment.

Environmental Monitoring and Assessment Program

The Agroecosystem component of the Environmental Monitoring and Assessment Program (EMAP) is developing a systems-level approach to monitor the "sustainability" of agroecosystems on a long-term, regional and/or national scale. The Agroecosystem Resource Group (ARG) is one of seven ecosystem resource components of the larger EMAP, established as an initiative of the U.S. Environmental Protection Agency (EPA) in 1988 (Kutz and Linthurst 1990). The ARG program is a cooperative program between EPA, the U.S. Department of Agriculture (USDA) (Heck et al. 1991), and North Carolina State University. One objective of the program is to provide current estimates of the condition of U.S. agroecosystem resources as a baseline against which future changes could be compared with statistical confidence. Through time, measurements for a suite of indicators and assessment endpoints will be taken from area segments selected from a probability sampling frame in collaboration with the USDA's National Agricultural Statistics Service (NASS). Indicators are being selected based on (1) the availability of techniques for obtaining measurements; (2) suitability of indicator for use in a single sampling period; and (3) interpretability of data (Meyer et al. 1992). Selected indicators and assessment endpoints are organized to reach the program's ultimate goal of developing an overall index of agroecosystem health or sustainability.

The ARG program has adopted the following definition of a healthy agroecosystem: "one that balances crop and livestock productivity with the maintenance of air, soil and water integrity and assures the diversity of wildlife and vegetation in the noncrop habitats" (Heck et al. 1991). This definition resembles definitions of sustainable agriculture.

The ARG is developing a number of indicators that address the three principle components of a sustainable agroecosystem: plant and animal productivity, environmental and ecological soundness, and socioeconomic viability. The initial pilot program being developed by the ARG will include indicators associated with the following assessment endpoints: crop productivity, soil quality, chemical use and export, water quality, and land use. Additional indicators are being developed to address assessment endpoints of animal production, quality of wildlife habitat, and socioeconomic viability. More specifically, ecological soundness and environmental quality will be measured by indicators of (1) soil integrity including soil structure, water and nutrient-holding capacity, vulnerability to erosion, extent of acidification, salinization and contamination, and nematode community patterns; (2) chemical export from fields; (3) irrigation water availability, quality, and runoff; (4) wildlife habitat quality; and (5) land use patterns. The efficiency of production will be measured by an aggregation of input indicators such as farm labor, mechanical power and machinery, agricultural chemicals, and seed purchases; and output indicators including crop production available for human or livestock consumption. Specific methods for monitoring socioeconomic viability are being developed and will include factors such as operator age, land tenure of individual fields, and management practices. Some indicators and assessment endpoints can be included under more than one component of sustainability.

The ARG program is in a developmental phase and welcomes innovative ideas for monitoring and assessing the ecological condition of agroecosystems. The program is designed as a vehicle for monitoring agroecosystem health. The program's monitoring efforts need support from fundamental, systems-level research of agroecosystem structure and function to help identify, evaluate, and interpret indicators that consistently measure aspects of sustainability.

REFERENCES

Altieri, M. A. 1987. *Agroecology: The scientific basis of alternative agriculture.* Boulder, CO: Westview Press. 227 pp.

Bidwell, O. W. 1986. Where do we stand on sustainable agriculture? *J. Soil Water Conserv.* 41:317-20.

Coleman, D. C., C. P. P. Reid, and C. V. Cole. 1983. Biological strategies of nutrient cycling in soil systems. *Adv. Ecol. Res.* 13:1-55.

Edwards, C. A. 1990. The importance of integration in sustainable agricultural systems. In *Sustainable agricultural systems,* ed. C. A. Edwards, L. Rattan, P. Madden, R. H. Miller, and G. House, 249-64. Ankeny, IA: Soil and Water Conservation Society.

Freckman, D. W. 1988. Bacterivorous nematodes and organic-matter decomposition. *Agric. Ecosys. Environ.*

Gliessman, S. R. 1990. Quantifying the agroecological component of sustainable agriculture: A goal. In *Agroecology: Researching the ecological basis for sustainable agriculture,* ed. S. R. Gliessman, 366-70. New York: Springer-Verlag.

Hauptli, H., D. Katz, B.R. Thomas, and R. M. Goodman. 1990. Biotechnology and crop breeding for sustainable agriculture. In *Sustainable agricultural systems,* ed. C. A. Edwards, L. Rattan, P. Madden, R.H. Miller, and G. House, 141-56. Ankeny, IA: Soil and Water Conservation Society.

Heck, W. A., C. L. Campbell, G. R. Hess, J. R. Meyer, T. J. Moser, S. L. Peck, J. O. Rawlings, and A. L. Finkner. 1991. *Environmental Monitoring and Assessment Program (EMAP)–agroecosystem monitoring and research strategy.* EPA/600/4-91. Washington, D.C.: U.S. Environmental Protection Agency.

Hendrix, P. F., D. A. Crossley, Jr., J. M. Blair, and D. C. Coleman. 1990. Soil biota as components of sustainable agroecosystems. In *Sustainable agricultural systems,* ed. C. A. Edwards, L. Rattan, P. Madden, R. H. Miller, and G. House, 637-54. Ankeny, IA: Soil and Water Conservation Society.

Ikerd, J. E. 1990. Agriculture's search for sustainability and profitability. *J. Soil Water Conser.* 45:18-23.

King, L. D. 1990. Soil nutrient management in the United States. In *sustainable agricultural systems,* ed. C. A. Edwards, L. Rattan, P. Madden, R. H. Miller, and G. House, 89-122. Ankeny, IA. Soil and Water Conservation Society.

Kutz, F. W., and R. A. Linthurst. 1990. A systems-level approach to environmental assessment. *Toxicol. Environ. Chem.* 28:105-14.

Lockeretz, W. 1988. Open questions in sustainable agriculture. *Am. J. Altern. Agric.* 3:174-80.

Luna, J. M., and G. J. House. 1990. Pest management in sustainable agricultural systems. In *Sustainable agricultural systems,* ed. C. A. Edwards, L. Rattan, P. Madden, R. H. Miller, and G. House, 157-73. Ankeny, IA: Soil and Water Conservation Society.

Madden, J. P. 1989. What is alternative agriculture? *Am. J. Altern. Agric.* 4:32-34.

Meyer, J. R., C. L. Campbell, T. J. Moser, G. R. Hess, J. O. Rawlings, S. Peck, and W. W. Heck. 1992. Assessing the ecological condition of U.S. agroecosystems.

In *Proceedings of the international symposium on ecological indicators,* October 16-19, 1990, Fort Lauderdale, Florida, Vol. I, ed. D. H. McKenzie, D. E. Hyatt, and V. J. McDonald. Essex, UK: Elsevier (in press).

Olson, G., and R. P. Breckenridge. 1992. Assessing sustainability of agroecosystems: An integrated approach. In *Proceedings of the international symposium on ecological indicators,* October 16-19, 1990, Fort Lauderdale, Florida, Vol. I, ed. D. H. McKenzie, D. E. Hyatt, and V. J. McDonald. Essex, UK: Elsevier (in press).

Parr, J. F., B. A. Stewart, S. B. Hornick, and R. P. Singh. 1990. Improving the sustainability of dryland farming systems: A global perspective. In *Advances in soil science,* vol. 13, *Dryland agriculture: Strategies for sustainability,* ed. R. P. Singh, J. F. Parr, and B. A. Stewart, 1-8. New York: Springer-Verlag.

Reganold, J. P., R. I. Papendick, and J. F. Parr. 1990. Sustainable agriculture. *Sci. Am.* 262:112-20.

Rodale, R. 1990. Sustainability: An opportunity for leadership. In *Sustainable agricultural systems,* ed. C. A. Edwards, L. Rattan, P. Madden, R. H. Miller, and G. House, 77-86. Ankeny, IA: Soil and Water Conservation Society.

Schaller, N. 1990. Mainstreaming low-input agriculture. *J. Soil Water Conserv.* 45:9-12.

Scofield, A. M. 1986. Editorial: Organic farming–the origin of the name. *Biol. Agric. Hortic.* 4:1-5.

Stenholm, C. W., and D. B. Waggoner. 1990. Low-input, sustainable agriculture: Myth or method? *J. Soil Water Conserv.* 45:13-17.

Wasilewska, L. 1979. The structure and function of soil nematode communities in natural ecosystems and agrocenoses. *Polish Ecol. Stud.* 5:97-145.

Using Knowledge of Soil Nutrient Cycling Processes to Design Sustainable Agriculture

P. F. Hendrix
D. C. Coleman
D. A. Crossley, Jr.

SUMMARY. Perspectives from different but complementary schools of thought in ecology are being applied to sustainable agriculture–population and community ecology on one hand and ecosystem ecology on the other. These perspectives intersect in the study of mechanisms and controls of processes in ecosystems. Increased understanding of these processes in agroecosystems will contribute to development of sustainable agriculture.

Current models of nutrient cycling in soil suggest that active or readily mineralizable fractions of soil organic matter (SOM) are coupled to plant-available nutrient pools through the processes of mineralization and immobilization. Factors that control these processes (soil temperature, water, and texture; and resource quality of organic inputs) are influenced by management practices and represent potential controlling points for managing nutrient cycles in agroecosystems.

Organic inputs present different problems for nutrient management than do synthetic fertilizers, because most nutrients must be mineralized before they are available to plants. This extra link in the nutrient flow

---

The authors are affiliated with the Institute of Ecology, University of Georgia, Athens, GA 30602. P. F. Hendrix is also affiliated with the Department of Agronomy, and D.C. Coleman and D.A. Crossley, Jr. with the Department of Entomology, University of Georgia, Athens, GA 30602.

This work was supported by grant BSR-8818302 from the National Science Foundation and a subcontract of grant 58-43YK-0-0045 from the U.S. Department of Agriculture to the University of Georgia Research Foundation.

The authors would like to thank Kathleen Dapkus for technical assistance and two anonymous reviewers for helpful comments.

pathway adds uncertainty to nutrient management. However, this may reflect more our lack of understanding of processes involved than an inherent unmanageability of organic fertilizers.

Further research is needed to determine longterm trends in organic-matter processes in agroecosystems; to improve predictive capabilities of nutrient-cycling process models and linkages with landscape-scale models; and to determine relationships between biodiversity of soil biota and nutrient cycling processes in agroecosystems.

INTRODUCTION

Foundations for sustainable agriculture are being built within a variety of disciplines. In ecology we are seeing the application of ideas from different but complementary schools of thought–population and community ecology on one hand and ecosystem ecology on the other. The former deal with interactions among organisms, populations, and their physical environment, and includes such topics as competition, predator-prey relationships, succession, and food web theory. Weed-crop ecology and integrated pest management are familiar examples (e.g., Altieri 1987; Vandermeer 1989; Gliessman 1990; Carroll et al. 1990). Ecosystem ecology is concerned principally with biogeochemistry and energetics of ecosystems, and with processes, such as primary production and decomposition, involved in fluxes of matter and energy through ecosystem components. Applications to agriculture include nutrient budget and nutrient cycling process studies (e.g., Frissel 1977; Lowrance et al. 1984; Stinner et al. 1984; Coleman and Hendrix 1988; Elliott and Cole 1989; Paul and Robertson 1989).

These different ecological perspectives intersect in the study of mechanisms and controls of processes in ecosystems, through their integration of biological, chemical, and physical phenomena (Figure 1). For example, microarthropods grazing on soil fungal populations can alter fungal activity which, in turn, alters plant residue decomposition rates in agroecosystems. These interactions are enhanced under conditions created by no-tillage management (Beare et al. 1992). Such situations are where ecology offers potential for the solution of contemporary problems in agriculture, i.e., a better understanding of how ecosystem processes operate in agroecosystems. Although we can apply knowledge derived from natural ecosystems to agricultural problems, we are not seeking only to mimic natural ecosystems in the design of sustainable agriculture. Rather, we must understand how agricultural management alters ecosystem processes and how management practices can be altered to take advantage of ecosystem processes.

FIGURE 1. The intersection of population/community ecology with ecosystem ecology. An important area of complementarity is the explanation of mechanisms and controls of ecosystem processes.

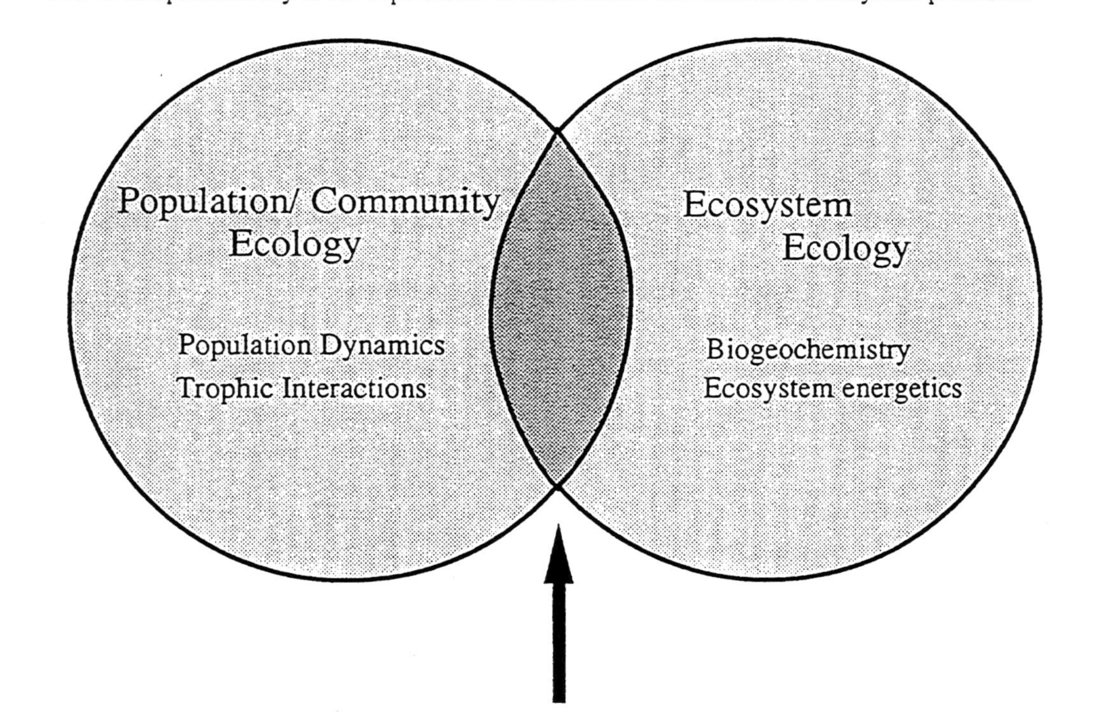

Three classes of ecosystem processes are particularly relevant to agricultural sustainability and environmental quality: (1) those involved in nutrient transformations (e.g., mineralization/immobilization); (2) those affecting soil structure and hydrology (aggregation and pore formation); and (3) those involving biological interactions (e.g., trophic relations within food webs). These processes are strongly influenced by managerial inputs into agroecosystems (e.g., fertilizers, soil tillage, and pest control) and may provide mechanisms for mitigating environmental degradation resulting from those inputs (e.g., eutrophication and groundwater pollution, erosion and sedimentation, pesticide residue contamination). Therefore, improved understanding of these processes in agroecosystems should help in the development of sustainable management practices and in enhancing environmental quality. This paper focuses on processes influencing nutrient transformations and soil structure, and their relationship to nutrient management in agroecosystems.

NUTRIENT CYCLING AND SOIL PROCESSES

Soil is a key component regulating nutrient cycles in terrestrial ecosystems. Soil serves as a reservoir that receives, stores, and releases most essential nutrients through an array of physical, chemical, and biological processes (weathering, fixation, volatilization, leaching, mineralization, etc.). Both organic and inorganic forms of nutrients are involved in these processes. In natural ecosystems, sizeable fractions of nutrient standing stocks may be present in living and dead organic matter. In conventional agroecosystems, particularly those under annual cropping, harvestable organic matter is removed and soil organic matter (SOM) may be reduced through accelerated decomposition. These losses, combined with inorganic fertilizer inputs needed to replace the lost nutrients, may result in lower overall size of nutrient standing stocks, lower proportions of nutrients in organic matter, and higher proportions of nutrients in mineral form in the soil. In humid regions, mineral nutrients (especially nitrogen) are susceptible to loss through leaching, soil erosion, and volatilization.

Thus, compared to nutrient cycles in natural ecosystems, those in conventional agroecosystems may be characterized by higher throughflow, lower storage capacity, and less recycling. Higher throughflow is a necessary condition for agroecosystems because of harvest outputs and required nutrient inputs. However, nutrient storage capacity and recycling can probably be increased to some extent in most agroecosystems, thereby increasing nutrient use efficiency, and sustainability. It is clear that organic matter plays a central role in nutrient storage capacity and nutrient recycling. Therefore, it is important that we understand the factors that regulate organic matter dynamics in agroecosystems in order to more effectively manage nutrient cycles.

Nutrient Transformations

In the first and still the most comprehensive study of its kind, Frissel (1977) presented a broad-scale evaluation of nutrient cycling in agroecosystems. The organizational model is simple in structure, containing only plant, livestock, and soil subsystems, but identifies 31 pathways along which nutrients flow in agroecosystems (Figure 2; Table 1). These include systems inputs and outputs and intrasystem transfers among nutrient pools. The pathways represent biological and chemical transformation of nutrients as well as physical transport mechanisms. Controls on these processes are also discussed, emphasizing the influence of management practices on nutrient loss and use efficiency.

The soil subsystem in Frissel (1977) shows the importance of the soil organic fraction in regulating nutrient availability to plants. Several more detailed models of nutrient cycling in soil organic matter have also been developed (Van Veen et al. 1984; Jenkinson et al. 1987; Parton et al. 1987). These include pools of organic matter of varying degrees of recalcitrance and thus different turnover and nutrient release rates. The plant-available nutrient pool is tightly coupled with the active, easily mineralizable labile or microbial biomass pool in the various models, through the key processes of mineralization and immobilization.

The Century model of Parton et al. (1987) is currently receiving wide interest in soil nutrient cycling research (Figure 3). This model has been used successfully to simulate SOM dynamics under various conditions, and has stimulated further development of concepts and methods in soil nutrient cycling studies (Coleman et al. 1989). The model has been formulated for carbon, nitrogen, phosphorus, and sulfur. It consists of plant residue and subsequent decomposition product pools (structural and metabolic) that release nutrients into active, slow, and passive pools with respectively longer turnover times.

For nitrogen, phosphorus, and sulfur, these pools connect to a mineral or labile pool available for plant uptake or loss to sink pools via element-specific processes (e.g., volatilization or leaching for nitrogen, occlusion for phosphorus). For carbon, losses from these pools are through gaseous CO_2 flux and leaching of soluble carbon compounds. Primary controls on nutrient flux rates among these pools are temperature, soil water, soil texture, and

FIGURE 2. Conceptual model of compartments and major fluxes of nutrients in agroecosystems. See Table 1 for definition of nutrient fluxes (from Frissel 1977 with permission).

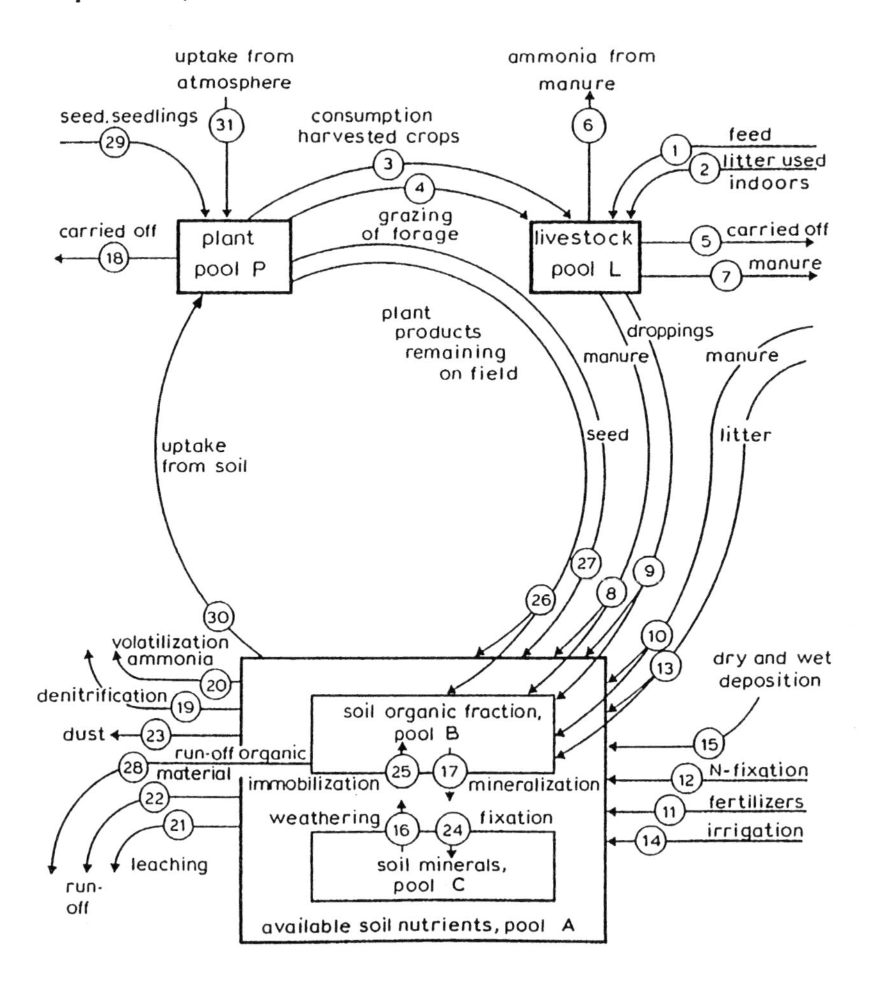

Table 1. Nutrient flux pathways important in agroecosystems. Numbers refer to Figure 2.

1. Input by feed for livestock
2. Input by litter used indoors
3. Transfer by consumption of harvested crops
4. Transfer by grazing of forage
5. Output of animal products
6. Output by losses from animal and/or manure in stables and/or feed lots (thus from any component before it reaches the soil); can be split into 6d, from droppings, and 6m, from manure
7. Output by manure (manure carried or sold)
8. Transfer by application of manure and/or waste (can be split into 8a or 8b)
9. Transfer by droppings on grazed areas (can be split into 9a and 9b)
10. Input by application of manure (can be split into 10a and 10b)
11. Input by application of fertilizer
12. Input by nitrogen fixation
13. Input by application of litter, sludge, and waste (can be split into 13a and 13b)
14. Input by irrigation, subsurface irrigation, or flooding (note: capillary rise is not included)
15. Input by dry and wet deposition (rain, dust, bird droppings) (Note: nutrients taken up directly by plants from rain or atmosphere are accounted for in item 31)
16. Transfer by weathering of soil mineral fraction
17. Transfer by mineralization of soil organic fraction
18. Output of primary products
19. Output by denitrification
20. Output by volatilization of ammonia (note: includes volatilization from manure, droppings, and fertilizers, thus from any material on or underneath the soil surface)
21. Output by leaching (note: net effect—leaching minus capillary rise)
22. Output of available nutrients by run-off (note: "Surface leaching" included)
23. Output by dust
24. Transfer by fixation in soil mineral fraction
25. Transfer by immobilization in soil organic fraction
26. Transfer by plant products (including litter) remaining on the field (can be split into 26a and 26b)
27. Transfer by seed for sowing
28. Output by run-off of organic matter
29. Input by seed or seedlings
30. Transfer by uptake of nutrients by the plant (can be split into 30t, top and 30r, root)
31. Input by nutrients taken up directly from atmosphere by plants

resource quality of the plant residue (e.g., carbon/nitrogen ratio, lignin or polyphenolic content). All of these controlling factors are strongly influenced by management practices and thus represent potential "entry points" for managing nutrient cycles in agroecosystems, as discussed below.

FIGURE 3. Nitrogen submodel of the Century model (from Parton et al. 1987 with permission).

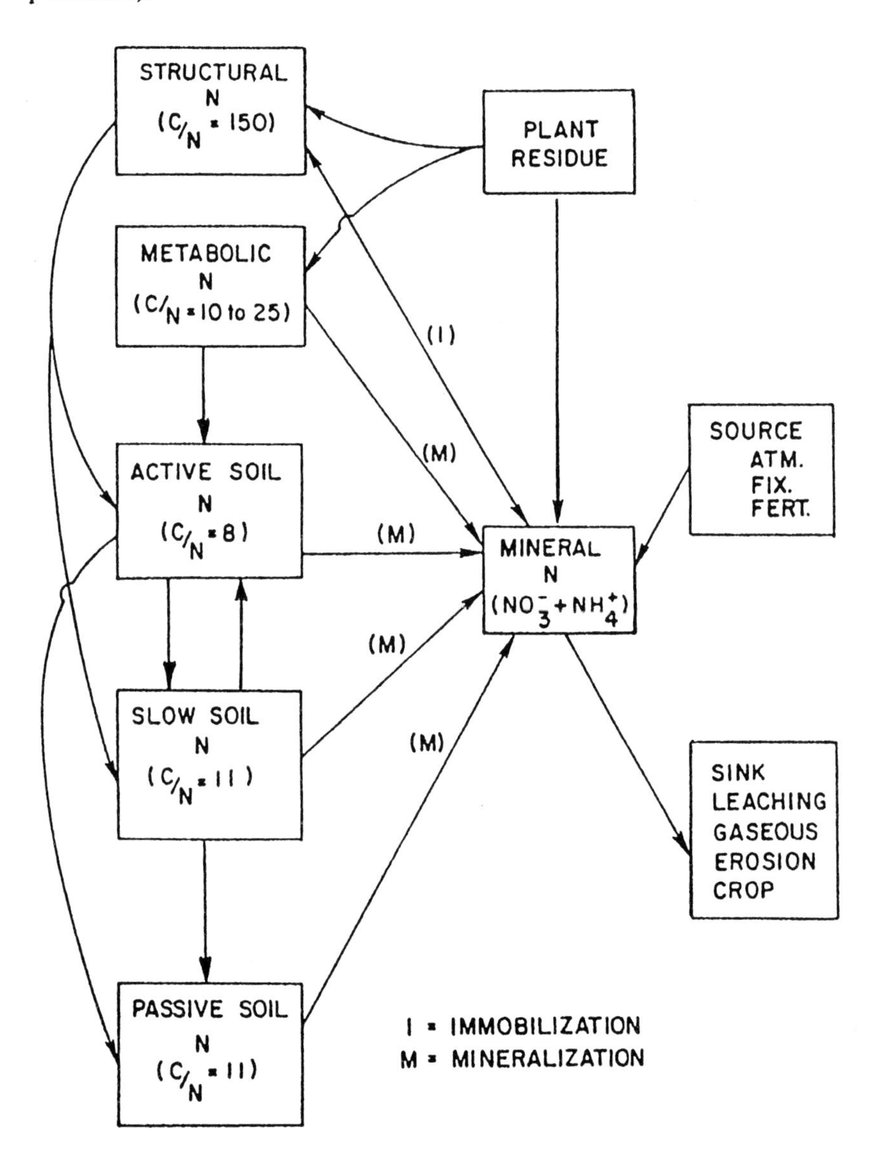

Soil Structure and Nutrient Dynamics

Biological, chemical, and physical processes in soil also affect nutrient cycling through influences on soil physical properties. Two mechanisms may be of particular importance in agricultural soils: formation of water-stable aggregates and formation of biopores. Soil aggregates are basic units of soil structure, composed of mineral and organic particles held together by a variety of forces (Boyle et al. 1989). Soil aggregates are formed through processes that can be viewed in a hierarchy of spatial scales (Tisdale and Oades 1982; Elliott and Coleman 1988). At fine scales, organic inputs from root exudates, plant residues, or organic amendments stimulate microbial production of polysaccharides and other compounds that bind mineral soil particles together into microaggregates. At coarser scales, macroaggregates are formed when fungal hyphae and fine roots entangle microaggregates and large mineral and organic particles, and when soil fauna (e.g., earthworms) produce fecal pellets or casts that consist of mixtures of mineral particles and organic materials of various sizes and in various stages of decay. The stability of these aggregates depends on soil physical and chemical characteristics, but their formation appears to be largely a function of biological activity in soils.

With respect to nutrient cycling, formation of water-stable aggregates represents a key ecosystem process through which organic matter may be protected from rapid decomposition and thus accumulate in soils. The actual protection mechanism appears to be encasement of SOM within aggregate chambers or "crypts," inaccessible to microbes (Boyle et al. 1989; Paul and Clarke 1989). Disruption of aggregates exposes protected SOM and results in a flush of microbial activity (Elliott 1986). This has important implications for nutrient management in agroecosystems. The natural cycle of aggregate formation and disruption via freezing/thawing or wetting/drying cycles and root proliferation may indirectly regulate the flux of nutrients between organic and inorganic pools. Management practices that disrupt soil aggregates (e.g., intensive tillage) or impede soil biological activity (e.g., use of certain chemicals) may tend to reduce organic content of soil over the long term.

A corollary process to aggregation is the formation of pores or spaces between aggregates. Well-aggregated soils have pore size distributions optimal for water and gaseous flux (Boyle et al. 1989). Biological activity may be a major factor in macropore formation. Roots and soil fauna penetrate soil, creating channels of sizes related to body diameter. Some macropores may be transient, being used once (e.g., insect emergence) and then infilled with soil matrix or surface material. Other macropores may be longer-lived, for example burrows of certain earthworm species that may be stable for long periods (Lee 1985). Alternate use of stable biopores by roots and earthworms may also occur. Wall linings of earthworm burrows are enriched with organic matter that helps stabilize burrow walls and serves as a site of increased microbial activity. Plowing disrupts macropores and alters overall pore size distribution.

The influence of porosity on nutrient cycling processes is a function of pore size. Water films within micropores provide suitable habitat for microflora and microfauna, such as bacteria, protozoa, and nematodes (Elliott et al. 1980; Elliott and Coleman 1988) and probably support a large fraction of biological activity in soil. The degree to which pores are filled with water influences relative proportions of aerobic versus anaerobic activity in soil. For example, 60 percent water-filled pore space (WFPS) is near optimal for aerobic activity (e.g., nitrification), while higher levels promote anaerobic activity such as denitrification. No-tillage management may promote denitrification by maintaining higher WFPS than in plowed soil, but significant N losses probably occur only in poorly drained soils (Doran and Smith 1987).

Macropores may affect nutrient cycling by influencing hydraulic conductivity and leaching of solutes such as nitrates. Interestingly, increased biological activity that accompanies reduced tillage (Hendrix et al. 1986, 1990) also increases macroporosity and water infiltration, which has led to concerns over possible nutrient leaching and groundwater contamination under no-tillage agroecosystems. High concentrations of nitrate have been observed in macropore solutions, but total amounts leached appear to be small (Edwards et al. 1989; Edwards 1991).

The models of soil nutrient dynamics and controlling factors presented in this section summarize our current understanding of key nutrient cycling processes in soil. These conceptual and analytical models offer a framework for devising nutrient management plans. The central role of SOM in soil nutrient dynamics suggests a logical focal point. In particular, the active or readily mineralizable nutrient pool and processes influencing it may be critical to effective nutrient management, as discussed below.

MANAGING NUTRIENT CYCLES

Several recent reviews have covered in detail basic and applied aspects of nutrient management strategies (Buchholz and Murphy 1987; Doran and Smith 1987; Heichel 1987; Parr and Colacicco 1987; Duxbury et al. 1989; Doran and Werner 1990; King 1990). This section will examine some of the nutrient cycling mechanisms discussed above with respect to their potential for management.

From a managerial view, controlling nutrient availability in soil and nutrient uptake by plants are the focus of a considerable amount of effort and expense in agriculture. Ineffective management is a primary cause of nutrient-related problems in environmental quality (e.g., eutrophication, groundwater contamination, atmospheric emissions). To improve sustainability of agriculture, the goal of nutrient management should be to achieve maximum nutrient uptake and recycling efficiency, to increase nutrient storage in active, available soil pools, and to minimize nutrient loss. However, different approaches and degrees of ecological understanding are required for inorganic versus organic nutrient inputs. Principal differences are in the point of entry of nutrients into the system (i.e., available versus unavailable pools) and predictability of their behavior.

Inorganic Fertilizer Management

A large body of literature and experience exists on the use of synthetic fertilizers in agriculture, and it need not be reviewed here. Inorganic fertilizers will undoubtedly remain an important source of nutrients in sustainable agriculture. Although they have not been used with great precision historically, techniques exist to achieve a high level of nutrient-use efficiency with these materials. Fertilizers usually are well characterized and tend to behave in relatively predictable ways compared with organic amendments.

Synchronization of nutrient supply and demand can be done through appropriate placement and timing of fertilizers relative to crop growth. In particular, concentrated placement near plant roots helps reduce chemical reactions with mineral soil (phosphorus fixation), enhances root uptake, "overwhelms" biological immobilization, decreases ammonia volatilization (Buchholz and Murphy 1987), and may reduce weed competition. Variable rate technologies, split applications, and slow release fertilizers in conjunction with soil testing can help maintain nutrient availability in quantities needed (Buchholz and Murphy 1987; Larson and Robert 1991).

Excess inorganic nitrogen (nitrate) in soil is highly susceptible to leaching and denitrification loss in humid climates. In reduced tillage systems, immobilization of nitrogen into organic matter and subsequent remineralization may provide a further slow-release mechanism if mineralization can be managed to coincide with plant growth.

Organic Matter Management

Until the middle of this century, organic matter has been the primary source of nutrients in agriculture. Renewed interest in organic matter management is creating the need for better understanding of the benefits and constraints of organic fertilizers and amendments. Of the organic materials used in U.S. agriculture, crop residues and animal manure constitute the largest proportions (Chen and Avnimelech 1986; Follett et al. 1987; SSSA 1987; King 1990).

Organic inputs present a different set of problems for nutrient management than do synthetic fertilizers. In contrast to the situation with mineral fertilizers, activities of soil biota are critical to the management of organic fertilizers because most nutrients must be mineralized before they are available to plants. This extra link in the nutrient flow pathway adds uncertainty to nutrient management. Manures, plant residues, and organic wastes tend to vary in nutrient and non-nutrient composition and to release nutrients in less predictable ways. This may reflect more our lack of understanding of the processes involved than an inherent unmanageability of organic fertilizers.

Plant material produced *in situ* provides the major organic input to most agroecosystems, although animal manures, organic wastes, and compost are gaining in importance. Inputs usually consist of by-products of crop production (e.g., crop and weed residues, roots), or green manures (e.g., legumes) produced specifically as organic inputs. Energy capture and carbon fixation via photosynthesis thus are the first steps in organic matter management (Bruce et al. 1991). Soil and climatic factors, particularly water availability, are principal determinants of primary production but management decisions must be made as to appropriate cropping systems and the partitioning of primary production between harvest and soil inputs. Harvest provides useable outputs, whereas inputs to soil are the basis for SOM formation and nutrient recycling.

Bruce et al. (1991) recognize two primary goals of organic matter management: (1) restoration and/or maintenance of SOM and surface physical properties via mulching and fueling of soil biota; and (2) nutrient enhancement for the benefit of subsequent crops. These goals are not mutually exclusive, but may require different approaches. Critical amounts of biomass required as soil inputs in either case depend on cropping systems, soil conditions, and temperature and water regimes. In general, to maintain SOM levels the input rate must at least equal the organic matter decomposition rate; to build SOM the input rate must be higher. Because some materials decompose more quickly than others, decisions must be made as to the type of material to apply. In systems where the primary goal is nutrient management, organic inputs should be used that will decompose and release nutrients synchronously with crop growth. For renovation of soils low in SOM, a combination of recalcitrant and labile organic materials may be required to protect and stabilize the soil surface while simultaneously beginning to fuel biological activity to increase aggregate stability, soil porosity, water infiltration, etc. (Doran and Smith 1987; Bruce et al. 1991). In both cases, knowledge of soil biotic activity patterns are needed.

Research in temperate agroecosystems has shown that organic matter inputs can increase SOM levels and soil fertility, depending on quantity and quality of the inputs (Doran and Werner 1990). Investigations in the subtropical southeastern United States show that SOM in highly degraded Ultisols also increases under management systems that enhance organic matter inputs. For example, Figure 4 shows chronosequences of percent carbon in the upper 15 cm of soil on the Georgia Piedmont under different crop cultures and initial conditions. The aggrading curve shows a trajectory of rapid carbon accumulation under sod-based management (alfalfa and fescue). Conversely, plowing and winter bare-fallow culture in organic-rich forest soils resulted in rapid losses of carbon, shown in the degrading curve. Together, these curves suggest lower and upper equilibrium levels for carbon storage in these soils, based on fertilizer-subsidized, *in situ* production of organic inputs. Most cropping systems in current use in this region probably approach equilibrium levels between these extremes.

As mentioned previously, the active fraction of the organic matter pool that accumulates over time appears to be the key to managing nutrient cycles in agroecosystems. This constitutes the pool of nutrients that can become available rapidly through mineralization and, conversely, that can build quickly through immobilization of inorganic nutrients. Management practices are needed that take advantage of these processes. A useful approach to sustainable management might be cultivation to stimulate mineralization during plant growth, residue return to immobilize excess inorganic nutrients, and continued organic inputs to replace nutrients removed in harvest.

Organic matter quality is an important consideration in the management of organic matter. Carbon and nitrogen (C/N), lignin, and polyphenolic content of plant material may strongly influence its decomposition rate (Coleman et al. 1989). Bruce et al. (1990) showed that despite significantly higher cumulative dry matter inputs over a 5-year period, soybean stover produced lower SOM levels and water-stable aggregate stability than did grain sorghum stover, both under spring disk-harrow, winter fallow crop culture. Higher nitrogen and lower lignin content in the soybean versus sorghum residue probably provided a higher-quality resource to soil biota, resulting in faster and more complete decomposition of the soybean organic inputs. Under no-tillage, double-cropping of grain sorghum plus crimson clover, residues produced 100 percent higher soil carbon and 50 percent

FIGURE 4. Chronosequences of carbon content in soils on the Georgia Piedmont.

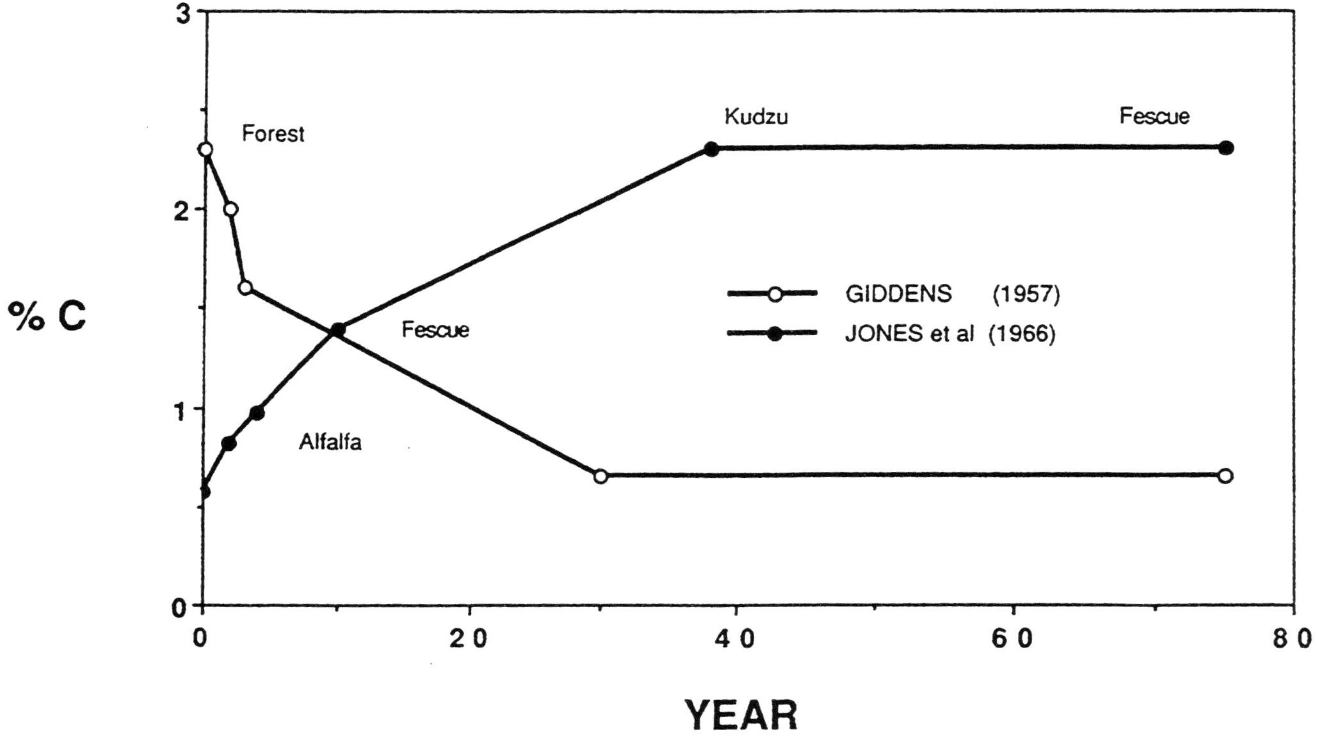

higher aggregate stability in comparison with the grain sorghum monoculture.

NUTRIENT CYCLING AND DIVERSITY OF SOIL BIOTA

Microflora and fauna of the soil are responsible for many aspects of nutrient transformation and storage. Maintenance of biodiversity of the soil biota may become a useful strategy for sustainable agriculture. Earthworms, for example, are known to influence nutrient cycling and soil structure. But a variety of earthworm species with potentially different effects on soil properties occur in soil (Lee 1985; Lavelle 1988). Deep burrowers may enhance macroporosity and thus improve water drainage through the soil profile. Shallow-burrowing, geophageous earthworms mix mineral soil and organic matter in the upper layers of the soil and may stimulate nutrient mineralization-immobilization processes. Litter dwellers may consume and speed decomposition of particulate organic matter on the soil surface. Thus, soil conditions that promote earthworm diversity (e.g., minimum tillage, high organic inputs, reduced pesticide use) may also enhance the effects of earthworms on soil properties.

Similarly, practices that favor biodiversity in general (e.g., polycultures, crop rotations, hedgerows, minimum tillage) may result in a number of benefits such as increased abundance of predators and beneficial parasites, and increased microhabitat diversity for microbial processes. But issues remain unresolved about diversity of soil biota and their impact on nutrient processes. The soil biota include a wide range of decomposers, predators, and parasites that impact the root system, decomposition systems, and structure of agricultural soils (Hunt et al. 1987). Protection of biodiversity, in itself, has been identified as an important goal (Lubchenko et al. 1991). In agricultural systems, we must ask (1) how does biodiversity vary among agricultural practices, (2) which practices may protect biodiversity, and (3) what is the relationship between species biodiversity and nutrient dynamics? The answers may not be obvious. For example, minimum tillage may protect biodiversity of oribatid mites, but another group, prostigmatid mites, showed increased species diversity under conventional tillage (Perdue 1987).

Currently, information about biodiversity of soil fauna in agroecosystems is being developed in such diverse studies as the Dutch Programme on Soil Ecology of Arable Farming Systems (Brussaard et al. 1988; Moore and de Ruiter 1991), the Swedish Ecology of Arable Land project (Andren et al. 1990), the Horseshoe Bend Agroecosystem project in Georgia, USA (Mueller et al. 1990; Beare et al. 1992), and others. Knowledge gained in these studies may provide guidance about the significance of biodiversity of soil biota for sustainable agriculture. At present, we point to the need for considerable basic and applied research in this area.

CONCLUSIONS AND FUTURE DIRECTIONS FOR RESEARCH

It is clear that an understanding of nutrient cycling processes in soil is necessary for development of sustainable agriculture. Current conceptual and analytical models suggest a number of options for managing nutrient cycles, but further research is needed on several factors:

1. An important focus for future basic research is long-term trends in organic matter dynamics under alternative management treatments. Ecosystem processes underlying these trends and cumulative effects on soil fertility and environmental quality should be determined from field investigations. Relative changes in soil nutrient pools–especially the active organic fraction–and the time required for soils to reach SOM equilibrium are particularly important. The emphasis here is on the long term; 5-10 or more years of continuous observation will be needed.
2. Continued development is needed of nutrient cycling process models and methods for more closely relating model components to measurable soil nutrient pools. Linkage of process models with other relevant types of models may be fruitful (Parton et al. 1989). Examples include explicit decomposer food web models whose dynamics may provide realistic controls on some processes being modeled (e.g., residue decomposition), and hydrologic models that may be critical in linking nutrient fluxes between ecosystems and across landscapes. At still larger scales, efforts to predict effects of agriculture on regional or global climate may benefit from linkages between atmospheric and soil process models. Long-term field experiments will be critical in validating and testing predictions from these models.
3. As mentioned previously, little is known about the relationships between biodiversity and ecosystem function, particularly in agroecosystems. Observational and experimental investigations in the field are needed to elucidate these relationships. Food web models may provide a good starting point for development of conceptual models.

Applied research needs also can be identified:

1. Management of organic inputs is less certain than that of conventional inorganic fertilizers. An applied goal of future research should be to make organic fertilizers as predictable and manageable as synthetic fertilizers (Sanchez et al. 1989).
2. There is growing demand for the organic matter produced in agroecosystems for uses other than harvest for food and animal forage (e.g., bioenergy fuels). Such exports of organic matter reduce inputs to soil and in many cases may increase the need for external inputs (i.e., fertilizer). Plant breeding and biotechnology programs may need to focus on the partitioning of total organic matter production rather than on yield; one goal may be to increase soil input relative to harvest export.

REFERENCES

Altieri, M. A. 1987. *Agroecology. The scientific basis of alternative agriculture.* Boulder, CO: Westview Press.

Andren, O., T. Lindberg, K. Paustian, and T. Rosswall. 1990. Ecology of arable land–organisms, carbon and nitrogen cycling. *Ecol. Bull.* 40:9-14.

Beare, M. H., R. W. Parmelee, P. F. Hendrix, W. Cheng, D. C. Coleman, and D. A. Crossley, Jr. 1992. Microbial and faunal interactions and effects on litter nitrogen and decomposition in agroecosystems. *Ecology,* in press.

Boyle, M., W. T. Frankenberger, Jr., and L. H. Stolzy. 1989. The influence of organic matter on soil aggregation and water infiltration. *J. Prod. Agric.* 2:290-99.

Bruce, R. R., G. W. Langdale, and L. T. West. 1990. Modification of soil characteristics of degraded soil surfaces by biomass input and tillage affecting soil water regime. *Trans. 14th Int. Conf. Soil Sci.* VI:4-9.

Bruce, R. R., G. W. Langdale, and P. F. Hendrix. 1991. The transfer of crop culture technology: Soil-climate dimensions. In review.

Brussaard, L., M. J. Kooistra, G. Lebbink, and J. A. Van Veen. 1988. The Dutch Programme on Soil Ecology of Arable Farming Systems I. Objective approach and some preliminary results. *Ecol. Bull.* 39:35-40.

Buchholz, D. D., and L. S. Murphy. 1987. Conservation of nutrients. In *Energy in plant nutrition and pest control,* ed. Z. R. Hesel, 101-103. Amsterdam: Elsevier.

Carroll, C. R., J. H. Vandermeer, and P. M. Rosset, eds. 1990. *Agroecology.* New York: McGraw-Hill Publishing Co.

Chen, Y., and Y. Avnimelech, eds. 1986. *The role of organic matter in modern agriculture.* Dordrecht, The Netherlands: Martinus Nijhoff Publishers.

Coleman, D. C., and P. F. Hendrix. 1988. Agroecosystem processes. In *Concepts of ecosystem ecology: A comparative view,* ed. L. R. Pomeroy and J. J. Alberts, 149-70. New York: Springer-Verlag.

Coleman, D. C., J. M. Oades, and G. Uehara. 1989. *Dynamics of soil organic matter in tropical ecosystems*. Honolulu: University of Hawaii Press.

Doran, J. W., and M. S. Smith. 1987. Organic matter management and utilization of soil and fertilizer nutrients. *Soil fertility and organic matter as critical components of production systems*, 53-72. SSSA Special Publication no. 19. Madison, WI: American Society of Agronomy Publishers.

Doran, J. W., and M. Werner. 1990. Management and soil biology. In *Sustainable agriculture in temperate zones*, ed. C. A. Francis, C. B. Flora, and L. D. King, 205-30. New York: John Wiley & Sons.

Duxbury, J. M., M. S. Smith, and J. W. Doran with C. Jordan, L. Scott, and E. Vance. 1989. Soil organic matter as a source and a sink of plant nutrients. In *Dynamics of soil organic matter in tropical ecosystems*, ed. D. C. Coleman, J. M. Oades, and G. Uehara, 33-67. Honolulu: University of Hawaii Press.

Edwards, W. M. 1991. Soil structure: Processes and management. In *Soil management for sustainability*, ed. R. Lal and F. J. Pierce, 7-14. Ankeny, IA: Soil and Water Conservation Society.

Edwards, W. M., M. J. Shipitalo, L. B. Owens, and L. D. Norton. 1989. Water and nitrate movement in earthworm burrows within long-term no-till cornfields. *J. Soil Water Conserv.* 44:240-43.

Elliott, E. T. 1986. Aggregate structure and carbon, nitrogen, and phosphorus in native and cultivated soils. *Soil Sci. Soc. Am.* J. 50:627-33.

Elliott, E. T., and C. V. Cole. 1989. A perspective on agroecosystem science. *Ecology* 70:1597-1602.

Elliott, E. T., and D. C. Coleman. 1988. Let the soil work for us. *Ecol. Bull.* 39:23-32.

Elliott, E. T., R. V. Anderson, D. C. Coleman, and C. V. Cole. 1980. Habitable pore space and microbial trophic interactions. *Oikos* 35:327-35.

Follett, R. F., S. C. Gupta, and P. G. Hunt. 1987. Conservation practices: Relation to the management of plant nutrients for crop production. In *Soil fertility and organic matter as critical components of production systems*, 19-51. SSSA Special Publication no. 19. Madison, WI: American Society of Agronomy Publishers.

Frissel, M. J. 1977. Cycling of mineral nutrients in agricultural ecosystems. *Agro-Ecosystems* 4:1-354.

Giddens, J. 1957. Rate of loss of carbon from Georgia soils. *Soil Sci. Soc. Proc.* 21:513-515.

Gliessman, S. R., ed. 1990. *Agroecology: Researching the ecological basis for sustainable agriculture*. New York: Springer-Verlag.

Heichel, G. H. 1987. Legume nitrogen: Symbiotic fixation and recovery by subsequent crops. In *Energy in plant nutrition and pest control*, ed. Z. R. Hesel, 101-31. Amsterdam, Elsevier.

Hendrix, P. F., R. W. Parmelee, D. A. Crossley, Jr., D. C. Coleman, E. P. Odum, and P. M. Groffman. 1986. Detritus food webs in conventional and no-tillage agroecosystems. *Bioscience* 36:374-80.

Hendrix, P. F., D. A. Crossley, Jr., J. M. Blair, and D. C. Coleman. 1990. Soil biota as components of sustainable agroecosystems. In *Sustainable agricultural sys-*

tems, ed. C. A. Edwards, R. Lal, P. Madden, R. H. Miller, and G. House, 637-54. Ankeny, IA: Soil and Water Conservation Society.

Hunt, H. W., and eight co-authors. 1987. The detrital food web in the shortgrass prairie. *Biol. Fert. Soils* 3:57-68.

Jenkinson, D. S., P. B. S. Hart, J. H. Rayner, and L. C. Parry. 1987. Modelling the turnover of organic matter in long-term experiments at Rothamsted. INTECOL Bull. 15:1-8.

Jones, L.S., O.E. Anderson, and S.V. Stavy. 1966. Some effects of sod-based rotations upon soil properties. *Georgia Ag. Expt. Stations Bull.*, No. 166, Athens, GA.

King, L. D. 1990. Sustainable soil fertility practices. In *Sustainable agriculture in temperate zones*, ed. C.A. Francis, C.B. Flora, and L.D. King, 144-77. New York: John Wiley & Sons.

Larson, W. E., and P. C. Robert. 1991. Farming by soil. In *Soil management for sustainability*, ed. R. Lal and F. J. Pierce, 103-12. Ankeny, IA: Soil and Water Conservation Society.

Lavelle, P. 1988. Earthworm activities and the soil system. *Biol. Fertil. Soils* 6:237-51.

Lee, K. E. 1985. *Earthworms: Their ecology and relationships with soils and land use*. Orlando, FL: Academic Press.

Lowrance, R., B. R. Stinner, and G. J. House, eds. 1984. *Agricultural ecosystems: Unifying concepts.* New York: John Wiley & Sons.

Lubchenko, J., and fifteen co-authors. 1991. The Sustainable Biosphere Initiative: An ecological research agenda. *Ecology* 72:371-412.

Moore, J. C., and P. C. de Ruiter. 1991. Temporal and spacial heterogeneity of trophic interactions with below-ground food webs. *Agric. Ecosyst. Environ.* 34:371-97.

Mueller, B. R., M. H. Beare, and D. A. Crossley, Jr. 1990. Soil mites in detrital food webs of conventional and no-tillage agroecosystems. *Pedobiologia* 34:389-401.

Parr, J. F., and D. Colacicco. 1987. Organic materials as alternative nutrient sources. In *Energy in plant nutrition and pest control*, ed. Z. R. Hesel, 81-99. New York: Elsevier.

Parton, W. J., D. S. Schimel, C. V. Cole, and D. S. Ojima. 1987. Analysis of factors controlling soil organic matter levels in the Great Plains grasslands. *Soil Sci. Soc. Am. J.* 51:1173-79.

Parton, W. J., R. L. Sanford, P. A. Sanchez, and J. W. B. Stewart with T. A. Bonde, D. A. Crossley, Jr., H. Van Veen, and R. Yost. 1989. Modeling soil organic matter dynamics in tropical soils. In *Dynamics of soil organic matter in tropical ecosystems*, ed. D. C. Coleman, J. M. Oades, and G. Uehara, 153-72. Honolulu: University of Hawaii Press.

Paul, E. A., and F. E. Clark. 1989. *Soil microbiology and biochemistry.* San Diego, CA: Academic Press.

Paul, E. A., and G. P. Robertson. 1989. Ecology and the agricultural sciences: A false dichotomy? *Ecology* 70:1594-97.

Perdue, J. C. 1987. Population dynamics of mites *(Acari)* in conventional and conservation tillage agroecosystems. Ph.D. dissertation, University of Georgia. 132 pp.

Sanchez, P. A., C. A. Palm, L. T. Szott, E. Cuevas, and R. Lal with J. H. Fownes, P. F. Hendrix, H. Ikawa, S. Jones, M. van Noordwizk, and G. Uehara. 1989. Organic input management in tropical agroecosystems. In *Dynamics of soil organic matter in tropical ecosystems*, ed. D. C. Coleman, J. M. Oades, and G. Uehara, 125-52. Honolulu: University of Hawaii Press.

Soil Science Society of America. 1987. *Soil fertility and organic matter as critical components of production systems*. SSSA Special Publication no. 19. Madison, WI: American Society of Agronomy Publishers.

Stinner, B. R., D. A. Crossley, E. P. Odum, and R. L. Todd. 1984. Nutrient budgets and internal cycling of N, P, K, Ca, and Mg in conventional tillage, no tillage, and old-field ecosystems on the Georgia piedmont. *Ecology* 65:354-69.

Tisdale, J. M., and J. M. Oades. 1982. Organic matter and water-stable aggregates in soils. *J. Soil Sci.* 33:141-61.

Vandermeer, J. 1989. *The ecology of intercropping*. New York: Cambridge University Press.

Van Veen, J. A., J. N. Ladd, and M. J. Frissel. 1984. Modelling C and N turnover through the microbial biomass in soil. *Plant Soil* 76:257-74.

Landscape Ecology: Designing Sustainable Agricultural Landscapes

Gary W. Barrett

SUMMARY. Landscape ecology focuses on the development and dynamics of spatial heterogeneity at large temporal/spatial scales. A landscape is a mosaic of elements (patches and corridors) generated at various scales. Agriculture has traditionally been managed at the agroecosystem (landscape patch) level of resolution and has been judged successful based on the concept of crop yield. A new transdisciplinary approach to agriculture at the landscape level needs to be implemented based on the concepts of sustainability, hierarchy theory, and landscape diversity. New transdisciplinary research and educational approaches must include problem-solving algorithms, the concept of net energy, and tools such as Geographic Information Systems. Recommendations are presented for changes that are urgently needed if sustainable agriculture is to be developed and managed at the landscape level. These recommendations include changes in and the integration of our present research, educational, and management philosophies. A new field of study–agrolandscape ecology–must evolve if we are to develop and manage agriculture in a sustainable and cost-effective manner for future generations.

INTRODUCTION

The need to integrate basic and applied science has long been recognized (e.g., McElroy 1977). Attempts have been made in the ecological sciences during the past decade to address this need (e.g., Institute of Ecology 1982;

Gary W. Barrett is Distinguished Professor of Ecology and Co-Director of the Ecology Research Center, Miami University, Oxford, OH 45056.

Franklin et al. 1990). Applied ecology has emerged as an integrative field of study that attempts to wed ecological theory with environmental application (Barrett 1984, 1987; Slobodkin 1988).

Ecology has frequently been defined as that scientific discipline which attempts to understand the structure, function, and behavior of natural ecological systems (Odum 1971; Johnson 1977). Applied ecology is defined as an interdisciplinary science that not only continues to advance ecological theory, but also attempts to evaluate the impact of humankind on the structure and function of natural, human-subsidized, and socioeconomic systems. Applied ecology also aids in the development of sustainable managed systems based on ecological theory.

Unless the science of applied ecology is based on a sound theoretical foundation, attempts to manage the environment are bound to fail (Lubchenco et al. 1991). A special report from the Ecological Society of America (ESA) entitled "The Sustainable Biosphere Initiative: An Ecological Research Agenda" (Lubchenco et al. 1991) outlines ecological problems at different temporal/spatial scales that demand an integrative perspective if solutions are to be found. Slobodkin (1988) noted that applied problems can guide ecology just as clinical problems have focused medical research. Indeed, the concept of preventative environmental management should underpin the management of sustainable ecological systems just as preventative medicine is the key to improving human health on a long-term basis.

During the past decade several new paradigms have emerged within the realm of applied ecology (Figure 1). A paradigm is a body of ideas that is sufficiently coherent to attract widespread acceptance (e.g., a new scientific discipline). The interrelationship of two of the paradigms–landscape ecology and agroecosystem ecology–constitutes the focus of this article and the framework for designing sustainable agricultural landscapes.

A landscape is a mosaic of elements (e.g., patches and corridors) generated at various scales. Natural and anthropogenic effects influence landscape pattern in a dynamic manner. Landscape ecology describes the development and dynamics of landscape patterns, the spatial and temporal interactions and exchanges of biotic and abiotic materials across the landscape, the influences of landscape patterns (i.e., spatial heterogeneity) on biotic and abiotic processes, and the management of this spatial heterogeneity for societal benefit and survival (Risser et al. 1984). Landscape ecology is motivated by the need to understand the development and dynamics of landscape patterns and how these patterns are related to ecological phenomena (Urban et al. 1987). An interdisciplinary approach is necessary to investigate these effects and interrelationships.

Earlier schemes attempting to depict levels of ecological organization

FIGURE 1. Diagram depicting the interface of applied ecology with emerging paradigms that integrate ecological theory with management practice (modified after Barrett, 1987, 1989b).

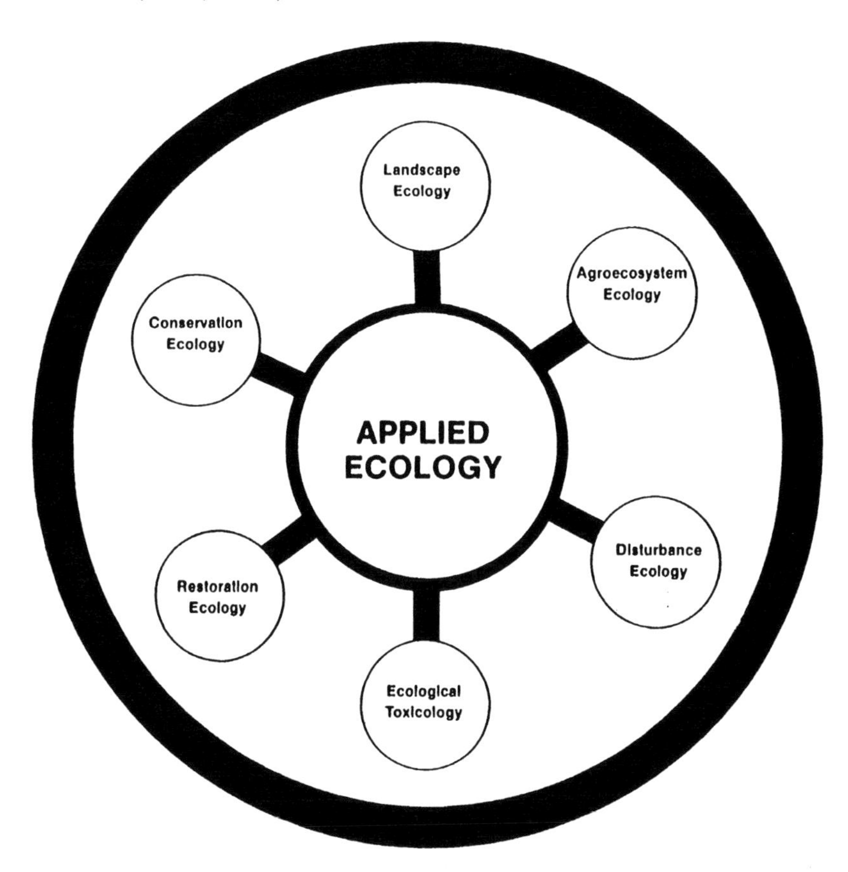

(e.g., MacMahon et al. 1978) omitted the landscape level of organization (Figure 2), although Rowe (1961) outlined a regional ecosystem level of integration. This omission limited the attention given to studies that addressed biotic and abiotic exchange rates (e.g., nutrient fluxes, rates of animal dispersal, and changes in biotic diversity). Understanding temporal and spatial exchanges across the landscape is critical when addressing problems related to the agricultural landscape mosaic (e.g., integrated pest management, cultural eutrophication, and biotic conservation).

Agroecosystem ecology is a relatively new field of study that weds ecological theory with practical application. The philosophy behind agroecosystem ecology is one of working with, rather than against, nature and of understanding that natural ecosystems represent excellent models for efficient and effective resource management systems (Barrett 1990). This philosophy includes the harmonious integration of individual agroecosystems within a total landscape perspective, including natural, subsidized, and socioeconomic systems (Altieri et al. 1983).

Although this paper will focus on the fields of landscape ecology and agroecosystem ecology (Figure 1), it must be recognized that all paradigms depicted are interrelated. For example, to understand agroecosystem-landscape interrelationships one also needs to understand the role of disturbance (Pickett and White 1985) as related to landscape heterogeneity; the importance of biotic diversity and connectivity (Allen and Starr 1982; Fahrig and Merriam 1985) as related to sustainability at the agroecosystem and landscape scales; and the impact on and management of nutrients and heavy metals (e.g., Levine et al. 1989) within and between landscape patches as related to ecological health and toxicology.

TRANSDISCIPLINARY THEORY AND CONCEPTS: A LANDSCAPE PERSPECTIVE

Several concepts and approaches must be considered and integrated when attempting to design sustainable agricultural landscapes (Figure 3). This section will discuss landscape concepts including sustainability, hierarchy theory, and landscape diversity. The following section will describe landscape approaches including problem-solving, net energy, and the use of Geographic Information Systems. This article will also illustrate how the concept of net energy can be used as an integrative management approach. Overall, there is a progression of integrative concepts and approaches from disciplinary reductionism to transdisciplinary holism (Jantsch 1972; Johnson 1977; Barrett 1984). Landscape concepts and approaches most frequently encompass interdisciplinary (coordination by higher-level concept) and transdisciplinary (multi-level coordination of entire system) theory and approaches (Jantsch 1972).

Sustainability

Sustain is defined as to keep in existence or to supply with necessities or nourishment to prevent from falling below a given threshold of health or

FIGURE 2. Levels of ecological integration, including the landscape level of organization.

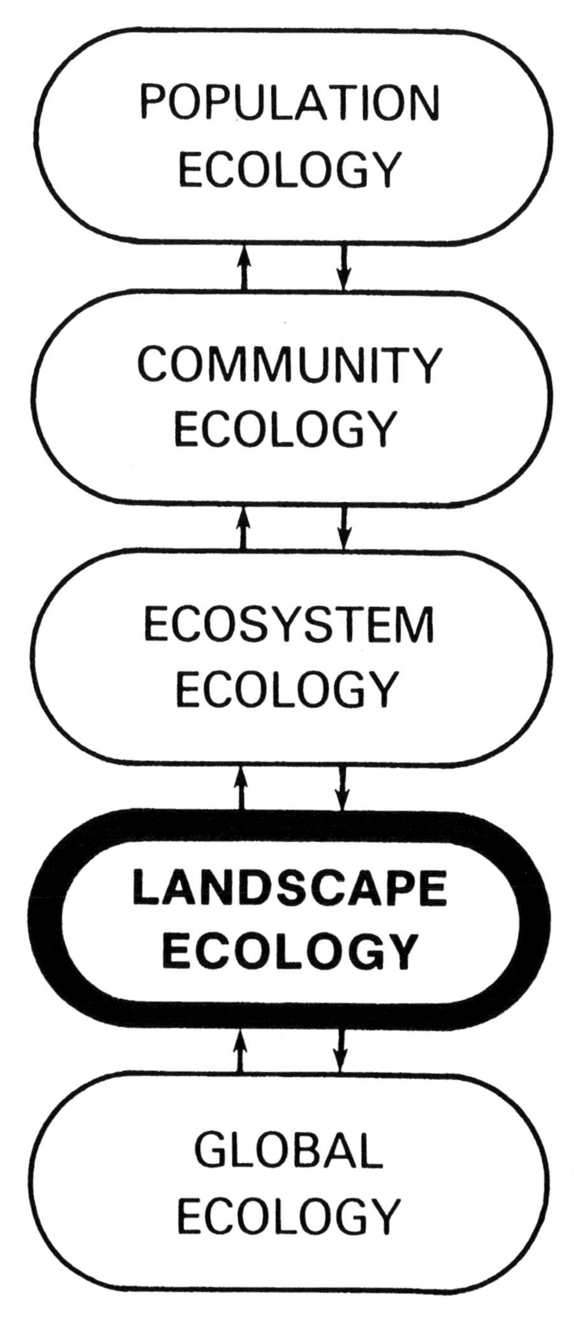

vitality. The term gained international stature when a United Nations report entitled *Our Common Future* (WCED 1987) popularized the concept of sustainable development which asserts that economic growth is essential to finding solutions to environmental problems. Numerous fields of study have accepted the concept of sustainability. For example, Gale and Cordray (1991) suggest eight approaches to the concept of sustainability in the field of forestry, including human benefit, "global village," and ecosystem-centered sustainability. Although the concept of sustained yield had long been at the center of forest management, these approaches suggest a broader definition of sustainability that encompasses ecosystem and global considerations. Interestingly, the landscape level of integration was not identified.

The recent report by the National Research Council of the National Academy of Sciences (NRC 1989) entitled *Alternative Agriculture* addressed the need to approach agricultural research from a perspective of reducing energy and chemical inputs, rather than from a strictly commodity-level crop yield perspective. Although the NRC report recognizes the need for interdisciplinary efforts to improve the understanding and management of agricultural systems, the need to develop a landscape perspective is poorly addressed.

The Ecological Society of America (ESA) in its Sustainable Biosphere Initiative (SBI) report (Lubchenco et al. 1991) referred to both sustainable forestry and sustainable agriculture, including needs and initiatives at the landscape level. The SBI report notes, however, that current research efforts are inadequate for dealing with sustainable systems that involve multiple resources, multiple ecosystems, and large spatial scales. Further, much of the current research focuses on commodity-based systems, with little attention paid to the sustainability of natural systems whose services currently lack a market value. Addressing the topic of sustainability at the landscape scale will require integration of social, physical, and biological sciences. The concept of sustainability nevertheless must become an integrative force when addressing these research and educational needs.

Hierarchy Theory

A hierarchical framework is necessary for research and management of agroecosystems at the landscape level (see O'Neill et al. 1986 for a review of hierarchy theory at the ecosystem level). Scientific research, including agroecosystem research, has all too frequently adopted a reductionist approach when a holistic or hierarchical approach should have been recognized. For example, problems frequently arise when pesticides are tested at one level (population or species) and then used without further study at

FIGURE 3. An integrative approach to research and management.

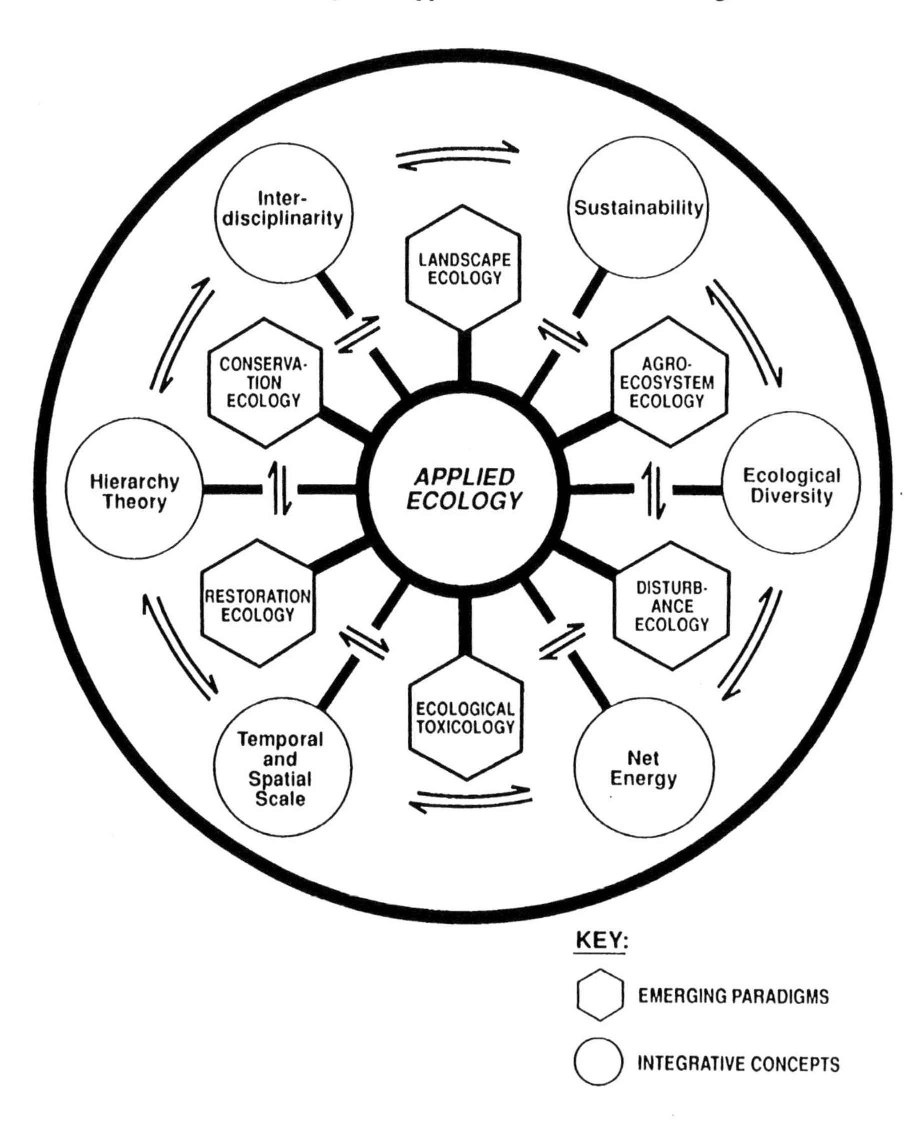

another level (ecosystem or landscape). Barrett (1988) provides an experimental design that illustrates a hierarchical research perspective.

There is also the hierarchy of processes (e.g., energy flow and regulatory mechanisms) at various trophic levels and at differing spatial/temporal scales. A hierarchical perspective helps to design research investigations at the most efficient and effective level of resolution.

Management of agricultural resources (e.g., crop yields) and problems (e.g., insect pest management) is frequently focused at the individual ecosystem (i.e., crop field) level. Long-term sustainable approaches, however, can best be implemented by management at the watershed or landscape scale because the flux of materials and organisms at these scales is the result of activities in the landscape mosaic. Therefore, proper management at the landscape level will influence decisions made at the individual field (Gould and Stinner 1984). Management at the landscape scale will require a more flexible policy and administrative approach. For example, cross-compliance of Federal conservation and farm programs (Steiner 1987; Myers 1988) could be used to manipulate landscape patterns and, hence, to promote (both ecological and economic) sustainable agricultural practices (Barrett et al. 1990). Hierarchy theory provides an integrative concept for addressing landscape patterns and heterogeneity that would be most beneficial both environmentally and economically. Unfortunately, there is a paucity of information at this scale. Lowrance et al. (1986) have outlined a hierarchical approach to sustainable agriculture and described interactions among various agronomic, economic, and ecological scales.

Agroecosystem research based on hierarchy theory needs to be conducted at a variety of spatial scales (Addicott et al. 1987; Urban et al. 1987). Long-term integrative research initiatives are needed (Callahan 1984; Barrett 1985), especially in the area of applied ecology that includes sustainable agroecosystem and landscape ecology (Barrett 1984; Barrett et al. 1990), because changes in ecosystem and landscape parameters are most frequently detectable in time scales measured in years.

Landscape Diversity

Numerous studies involving biotic (especially species) diversity have been conducted using alpha (within-habitat) and beta (between-habitat) diversity indices. There is a paucity of information, however, regarding the use of gamma (landscape-level) indices to describe large-scale patterns, processes, or phenomena. Not only is it necessary to measure the richness of patch (e.g., old-field, forest, and crop) types present in the landscape, but the degree of habitat connectivity (Fahrig and Merriam 1985) is also vital in describing landscape structure and function in an agricultural landscape mosaic. For example, Barrett et al. (1990) used the Shannon-Wiener index ($H' = -\Sigma pi \ln pi$) to measure the diversity of crops and to describe the agricultural landscape in Ohio for the years 1940 and 1982. They found that crop diversity decreased from .80 in 1940 to .60 in 1982. This decline in diversity was due mainly to a decline in the number of crops, especially small grains (e.g., rye and barley) and hay crops (e.g., clovers and timothy). Hay crops, which contribute significantly to soil development, have been dropped from almost all crop rotations in Ohio (Barrett et al. 1990). Less crop diversity may also result in greater crop losses to pests (Power 1987). Furthermore, less landscape diversity may mean the elimination of overwintering habitats for beneficial insects and the enhanced movement of pests between crop fields.

It is also important to increase diversity at the field (agroecosystem) scale. Corridors, for example, can be established within the crop matrix. Kemp and Barrett (1989) found that potato leafhoppers (*Empoasca fabae*) avoided grassy corridors planted in soybean agroecosystems and were less abundant in soybeans adjacent to grassy corridors, suggesting that these corridors confer an "associational resistance" (Altieri et al. 1977) to invasion by migrating adults. *Nomuraea rileyi*, a fungal pathogen, also infected a significantly higher proportion of green coverworms (*Plathypena scabra*) in soybean plots divided by grassy corridors. Interestingly, planted grassy corridors were found to be more effective in reducing insect defoliators than were successional corridors (Kemp and Barrett 1989). The authors concluded that grassy corridors represent an alternative to insecticide pest control management in soybean agroecosystems that merits further investigation.

At the larger agroecosystem field scale, Altieri et al. (1983) and Horwith (1985) outlined the economic and ecological benefits of multi- or intercropping as an alternative to "modern" monoculture agricultural practice. Multicropping also has the advantage of increasing patch diversity at the landscape level.

Barrett and Bohlen (1991) discuss the importance of landscape corridors (e.g., fencerows, hedgerows, and stream corridors) as major structural and functional components (elements) of the landscape that frequently connect landscape patches (e.g., agricultural fields). They note that corridors are increasingly recognized as important landscape elements for ecological regulatory processes such as (1) animal dispersal, (2) habitat for nongame species, (3) prevention of soil and wind erosion, and (4) integrated pest management. Thus, corridors are important in the conservation and regulation of biotic diversity at both the patch (field) and landscape levels of integration. Lubchenco et al. (1991) recommend that new research efforts address both the importance of biotic diversity in controlling ecological processes and the role that ecological processes play in shaping patterns of diversity at different scales of time and space. Thus, the concepts of sustainability, hierarchy theory, and biotic diversity are interrelated; a better understanding of these relationships will help scientists, resource managers, and policy makers to design sustainable agricultural landscapes for the future.

TRANSDISCIPLINARY APPROACHES: A LANDSCAPE PERSPECTIVE

Although there is an array of approaches, algorithms, and methodologies (e.g., cost-benefit analysis, cybernetics, and the scientific method) available for addressing problems at different scales, I will focus here on three methodologies and concepts (a problem-solving algorithm, net energy, and Geographic Information Systems) that will likely contribute most in helping to design sustainable agricultural landscapes during the coming decade.

Problem-Solving Approach

The need to restructure (i.e., to increase the number of landscape patches and corridors at the landscape level) will require that a problem-solving approach (Barrett 1985) be employed, since biological, physical, and socioeconomic parameters will need to be evaluated and integrated accordingly. Figure 4 depicts a problem-solving approach to resource management at the landscape scale. Barrett and Bohlen (1991) described how this algorithm could be applied to establish landscape corridors in an efficient and cost-effective manner. Although it has been shown that landscape patches need to be connected in order to provide dispersal routes for birds and small mammals within an agricultural landscape mosaic (Wegner and Merriam 1979; Henderson et al. 1985; Lorenz and Barrett 1990), additional parameters must be evaluated (e.g., economic costs, spread of disturbance, erosion control, edge effects, and biotic diversity, to name a few) before a final solution is implemented. The problem-solving algorithm (Figure 4) provides a methodology that attempts to integrate biological, physical, and socioeconomic factors in a holistic manner (for details see Barrett 1985; Barrett and Bohlen 1991).

Previously (Barrett 1981, 1984, 1985), I have argued that the noösystem, rather than the ecosystem (Evans 1956), should be used as the basic unit of study for integrating biological, physical, and socioeconomic parameters within a holistic framework. The noösystem concept would include not only the study of the structure and function of natural systems, but would also embrace the social, economic, and cultural influences on ecological systems. A landscape perspective helps to ensure that theory and application are wed as an integrative process. Naveh and Lieberman (1984) call attention to the importance of noöspheric-cultural influences and impacts on resource

FIGURE 4. A 19-step problem-solving algorithm used for resource management purposes (modified after Barrett, 1985, Bioscience 35(7): 423-427. Copyright 1985 by the American Institute of Biological Studies).

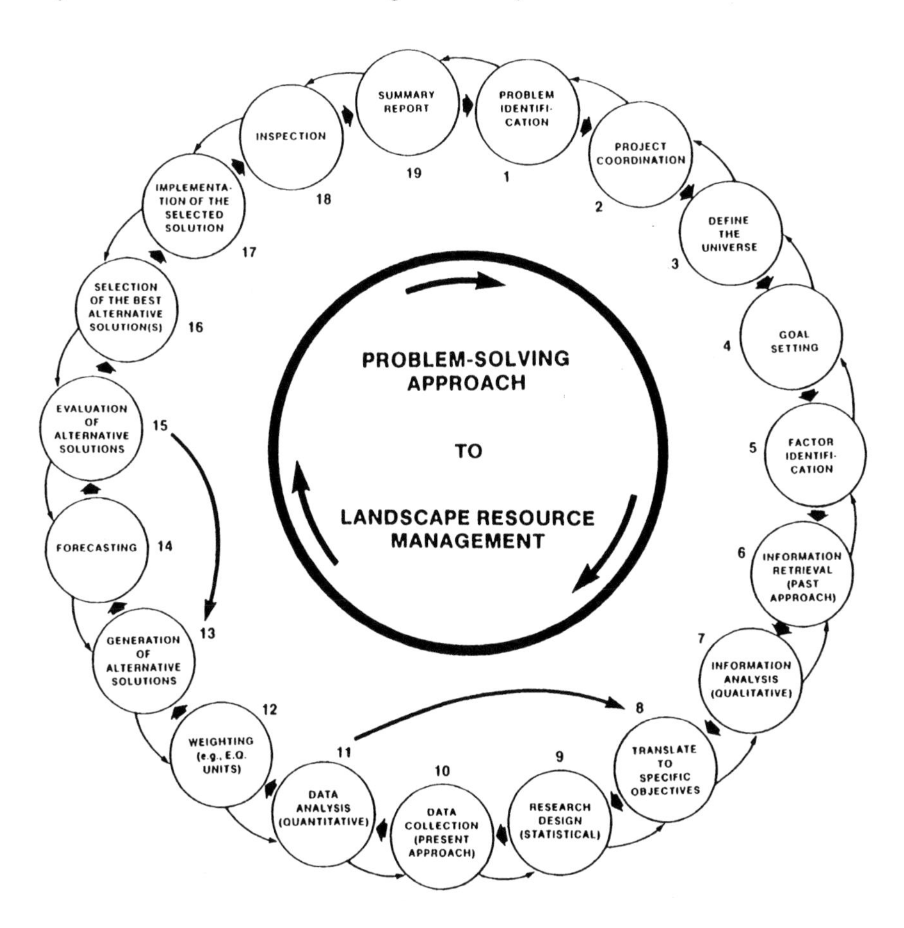

management in the field of landscape ecology. It should also be noted that the problem-solving algorithm is an interdisciplinary approach to resource management needed to implement a transdisciplinary research and educational philosophy based on sustainability–a philosophy necessary for societal benefit and survival.

Net Energy

A net energy approach to "ecological bookkeeping" is depicted in Figure 5. Net energy is defined here as the net primary productivity yield (A) from an energy production system (i.e., from an agroecosystem over an annual cycle) minus the amount of energy subsidies (B) needed to sustain that yield. Energy subsidies to the food system include (1) fossil fuels used for planting, cultivating, harvesting, and transporting the crop; (2) fertilizers used to maintain or restore nutrient availability to the system; and (3) pesticides used to help regulate or destroy weeds, insects, or soil organisms thought to be harmful to the system. In the United States approximately ten units of energy are put into the food system as subsidies only to get out one unit of food energy (Steinhart and Steinhart 1974). A natural ecosystem, by comparison, requires no subsidies to sustain its yield. That is why it is important that one understand the theory and behavior of natural, unsubsidized, solar-powered ecosystems and that they serve as model systems (Altieri 1987; Barrett 1990; Barrett et al. 1990; Lubchenco et al. 1991) in relation to the development of a sustainable approach to agroecosystem management.

How agroecosystems relate to or can be linked with natural ecosystems (e.g., old-fields or forests) to establish a lower-input sustainable landscape (LISL) also needs to be addressed. Insect pests such as the Mexican bean beetle (*Epilachna varivestis*), for example, frequently overwinter among plant litter in forest patches, but forage extensively in soybean crops during summer months. It has also been shown that, in the course of flight, herbivores occasionally wander away from patches or crop fields in their search for food (Kareiva 1983). Markov models (see Kareiva [1982] for mathematical background in a landscape setting) help to interpret experimental results and predict the response of pests to novel cropping manipulations or patterns. In fact, Turchin (1986) used Markov modeling to predict the effects of host patch size on the movement of the Mexican bean beetle. Thus, insect dispersal, movement, and reproductive patterns can often best be investigated at the landscape scale rather than strictly at the field scale. Shoemaker (1981) and Kareiva (1986) noted that integrated pest management (IPM) theory frequently ignored both the spatial component of pest population dynamics and the contributions of pest movement to patterns of crop dam-

FIGURE 5. A net energy approach to agroecosystem management. See text for details.

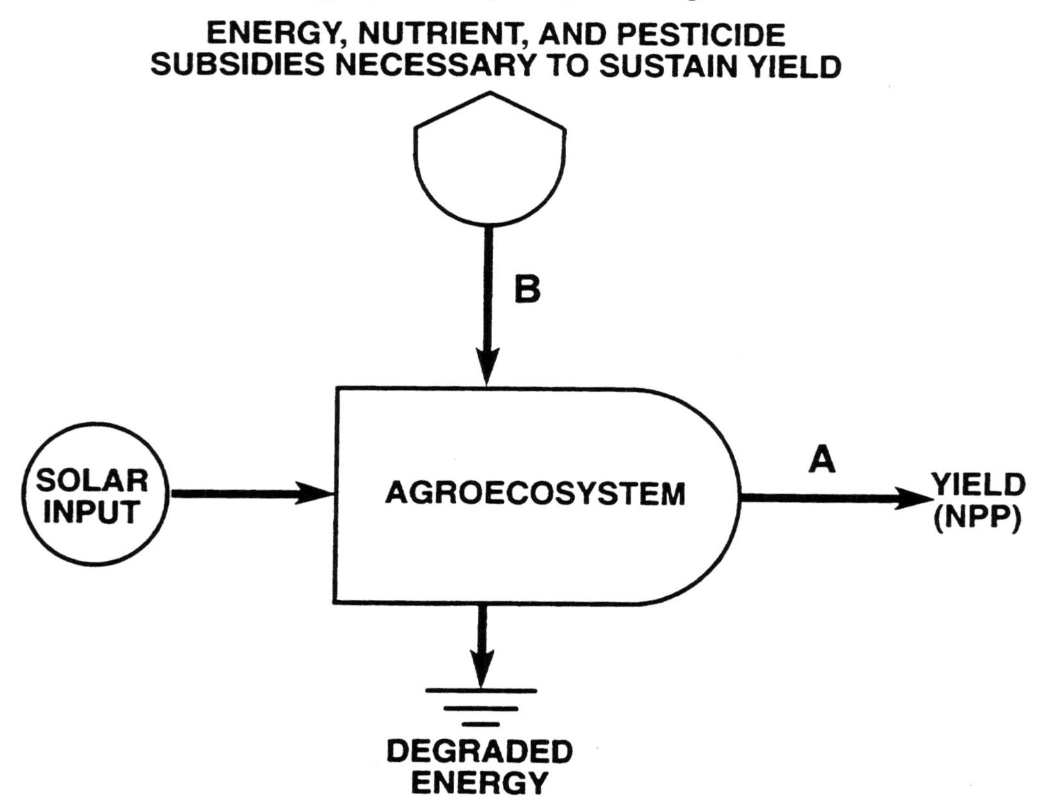

age. Future IPM strategies to control insect pests should focus on the manipulation of landscape elements rather than on increased use of pesticide subsidies at the field scale.

Nutrients should also be recycled within field patches and watersheds as a substitute for reliance on increased nutrient inputs. Well-managed alternative farming systems nearly always use less synthetic chemical pesticides, fertilizers, and antibiotics per unit of production than comparable conventional farms. Reduced use of these inputs lowers production costs and lessens agriculture's potential for adverse environmental and health effects without necessarily decreasing–and in some cases increasing–per unit crop yields (NRC 1989). It is imperative that the ratio of energy inputs to energy yields (Figure 5) be decreased dramatically if we are to design healthy and sustainable rural landscapes. An assessment of actual net energy inputs (especially subsidies) and outputs (benefits in addition to yield) will educate farmers, resource managers, and policy makers regarding the design of a sustainable landscape–a landscape that must increasingly be managed at a greater spatial scale.

Geographic Information Systems

Inventory of land-use practices and reconnaissance of resources for large regions (e.g., watersheds and landscapes) have resulted in the development of a variety of computer database systems. Geographic Information Systems (GIS) now serve as an inventory, analysis, and management tool for these landscape-level needs and management objectives. Unfortunately, much remains to be accomplished regarding an inventory of biotic resources at the regional scale. Recent developments in remote sensing and GIS technologies, however, now permit examination of ecological patterns at spatial scales larger than the agroecosystem level. At the same time, there is increased appreciation for the importance of processes (e.g., dispersal and recruitment) at small spatial scales (Lubchenco et al. 1991). Long-term ecological studies and the development of new monitoring techniques to reconstruct past communities and landscapes have also extended the temporal scale. In addition, it is important to note that temporal and spatial scales interact. For example, rare events such as the 1988 Yellowstone fires (Knight and Wallace 1989) may have profound effects on spatial patterns.

Both the continued collection of landscape-level data across temporal and spatial scales and the development of remote sensing technologies are urgently needed if society is to plan for and develop a sustainable landscape. There is now an urgency in solving large-scale environmental problems (e.g., spread of disease, water quality control, and biotic conservation); GIS is an increasingly important inventory, assessment, and management tool. An increased monitoring capability, such as the U.S. Environmental Protection Agency's proposed Environmental Monitoring and Assessment Program (EMAP) (Messer et al. 1991), would help to achieve these monitoring and assessment needs at a larger scale.

DISCUSSION: A FUTURISTIC PERSPECTIVE

Research and educational goals must change in the 1990s if a truly integrative and sustainable approach to agriculture is to be realized. Abelson (1990) noted that research activities in agriculture are destined to be fostered by a conjunction of factors. For example, new farm legislation will change provisions of existing laws that have traditionally subsidized a monoculture approach to crop production; new legislation will increase funds for competitive agricultural research grants; and societal and governmental pressures are mounting to reduce pollution through reduced use of fertilizer, pesticides, and fossil fuels. It is also recognized that the world's human population is continuing to increase as well as problems related to world food production, energy resource management, genetic diversity, and environmental contamination. As a result, a new transdisciplinary, holistic approach to agroecosystem ecology must emerge. In fact, the relatively new field of agroecosystem ecology must either evolve into or become a major component of a new paradigm which I will term agrolandscape ecology (i.e., the study of agroecosystems at the landscape scale–study that not only analyzes the interaction among crops and natural systems, but also the design and management of the total landscape based on the concept of sustainability).

Listed below are research, educational, and management recommendations that should be encompassed within this emerging field of study:

a. *Production systems should be managed from an input rather than an output perspective.* Odum (1989) noted that nonpoint sources of pollution can only be reduced and controlled by input management that involves increasing the efficiency of production systems and reducing the input of environmentally damaging materials. Input management, therefore, requires a major change in the approach to agriculture; this approach will focus on input reduction and recycling, rather than on waste disposal and environmental cleanup initiatives.

Michael R. Deland, Chair of the President's Council on Environmental Quality, recently stressed (Deland 1991) that "the time has come for a new course–to emphasize pollution prevention, not pollution control." Pimentel et al. (1991) noted that a substantial reduction in pesticide use might increase food costs only slightly. Thus the time has arrived to implement an input management approach to agrolandscape ecology.

b. *Efforts should be made to couple production systems with natural systems based on a landscape perspective.* Instead of viewing new fields of study such as urban ecology (i.e., a heterotrophic [P/R < 1] systems approach) and agroecosystem ecology (i.e., an autotrophic [P/R > 1] systems approach) as separate fields of study, let us immediately focus on how these fields of study might be coupled to provide for a sustainable landscape and a sustainable society (Barrett 1989a). This wedding will be one of the great challenges of the 1990s involving systems analysts, landscape ecologists, landscape architects, agroecosystem scientists, resource managers, and public policy makers; the economic and ecological benefits, however, will be enormous.

The coupling of natural with subsidized systems will also help to generate new approaches to

- integrated pest management based on natural regulatory mechanisms rather than on increased use of pesticides;
- water quality management based on total watersheds (i.e., a nonpoint pollution control approach) rather than on individual problems (i.e., a point source cleanup approach); and
- energy resource management based on a diversity of solar-based energy resources rather than on an increased consumption of finite fuels.

This integrative approach will also provide a better understanding regarding the role of landscape heterogeneity in the spread of disturbance, and the capacity for natural systems to aid feedback control and regulation of production systems (Barrett et al. 1990). As a positive spin-off of this approach, socioeconomic factors will be integrated into the decision-making process.

c. *Efforts should be made to establish new educational and research centers of excellence focused on new inter- or transdisciplinary fields of study that integrate the academic-industrial-governmental triumvirate.* Lines between basic and applied science are blurring. Disciplinary divisions are becoming irrelevant–even counterproductive–as many of the most striking research advances are occurring at the interface of multiple fields (Holden 1991). New fields of study that emerged in the 1980s such as landscape ecology, agroecosystem ecology, ecological toxicology, and restoration ecology represent interface paradigms that effectively integrate ecological theory with practical application. The Association of Ecosystem Research Centers (AERC) now contains 39 member centers that advance ecosystem science as an integrative problem-solving method at local, regional, national, and international levels. The Long-Term Ecological Research program (LTER) sponsored by the National Science Foundation (NSF) now includes 17 LTER sites large enough to incorporate moderate to large landscape mosaics (Franklin et al. 1990).

Unfortunately, the NSF LTER program has yet to network effectively with "satellite" centers in order to address large-scale applied problems at the landscape scale; an LTER or national site focused on both an educational and research agenda in the area of agrolandscape ecology at the national or biosphere scale is urgently needed. Such a site must include social, biological, and physical science personnel from academia, industry, and government whose purpose would be to design a comprehensive research and educational agenda for the twenty-first century. As recently stated in *Science* (Holden 1991) by physicist Donald Shapero of the National Academy of Science, "Academia is back in the 12th century in terms of organization." John White of NSF also wondered if "we are really doing science a service by narrow disciplinary organization at the agency." It has become increasingly clear that the answer to White's question is "no"; new organizational structures, centers, and institutes with proper reward systems (especially for young investigators) must be established if we are to solve complex societal problems concerning sustainable agricultural and landscape development.

d. *New institutional mechanisms for training, retooling, and educating personnel, including the public, in agrolandscape ecology must be established on a national basis.* Undergraduate and graduate students must be provided with courses, majors, and internship opportunities dealing with theory, problems, and management techniques at a greater temporal/spatial scale; the nontraditional student must become an integral part of the academic community; and the present workforce (e.g., farmers, resource managers, policy makers, extension agents, and agrochemists) must be provided with training and "retooling" programs that will not only reduce unemployment, but equally important, will help to solve major environmental problems at the local, regional, and national levels.

As discussed in the September, 1989 special issue of *Scientific American* (Crosson and Rosenberg 1989), agricultural science and technology may indeed find ways to feed 10 billion people in the future, but social and economic changes will need to be made to persuade farmers to adopt new technologies that will maintain food production without further degrading the environment. The authors note that in the long run the most successful approaches will rest on merging individual and societal interests. What is lacking are forms of communication that connect society's interests in a sustainable agricultural system with the well-being of the individual farmer. Specifically, institutional mechanisms must be established that engender an integrative approach to landscape-level resource management. Establishing these mechanisms has now become the most important challenge regarding a sustainable approach to agricultural and landscape development.

CONCLUSIONS

It is likely that the decade of the 1990s will be labeled the "Decade of Sustainable Development." Administrators, policymakers, educators, students, and the public at large must recognize, however, that this concept and management goal requires new transdisciplinary approaches to education, a new and integrative research agenda at greater temporal/spatial scales, governmental policies that encourage and reward industry for such ecological functions as recycling and energy conservation, and creative approaches to educating the public regarding the long-term economic and environmental benefits resulting from a sustainable approach to landscape management.

The development of a sustainable landscape, including the education of its citizenry, must include strategic planning, local and national leadership, transdisciplinary educational initiatives, advanced training and technologies, and public support if we as a nation are to manage vital resources in a cost-effective manner. Once we as a society recognize that sustainable development is in our best interests, both economically and ecologically, then it will be possible to design sustainable agricultural landscapes that will benefit future generations.

REFERENCES

Abelson, P. H. 1990. Opportunities in agricultural research. *Science* 248:941.

Addicott, J. F., J. M. Aho, M. F. Anotlin, D. K. Padilla, J. S. Richardson, and D. A. Soluk. 1987. Ecological neighborhoods: Scaling environmental patterns. *Oikos* 49:340-46.

Allen, T. F. H., and T. B. Starr. 1982. *Hierarchy: Perspectives for ecological complexity.* Chicago: University of Chicago Press.

Altieri, M. A., A. Schoonhoven, and J. D. Doll. 1977. The ecological role of weeds in insect pest management systems: A review illustrated with bean (*Phaseolus vulgaris*) cropping systems. *Proc. Natl. Acad. Sci. USA* 23:185-205.

Altieri, M. A., D. K. Letourneau, and J. R. Davis. 1983. Developing sustainable agroecosystems. *Bioscience* 33:45-49.

Altieri, M. A. 1987. *Agroecology: The scientific basis of alternative agriculture.* Boulder, CO: Westview Press.

Barrett, G. W. 1981. Stress ecology: An integrative approach. In *Stress effects on natural ecosystems*, ed. G. W. Barrett and R. Rosenberg, 3-12. New York: John Wiley & Sons.

Barrett, G. W. 1984. Applied ecology: An integrative paradigm for the 1980s. *Environ. Conserv.* 11:319-22.

Barrett, G. W. 1985. A problem-solving approach to resource management. *Bioscience* 35:423-27.

Barrett, G. W. 1987. Applied ecology at Miami University: An integrative approach. *Bull. Ecol. Soc. Am.* 68:154-55.

Barrett, G. W. 1988. Effects of Sevin on small-mammal populations in agricultural and old-field ecosystems. *J. Mammal.* 69:731-39.

Barrett, G. W. 1989a. A sustainable society. *Bioscience* 39:754.

Barrett, G. W. 1989b. Applied ecology: An emerging integrative paradigm for the 1990s. *Assoc. Systematics Collections Newslett* 17:37.

Barrett, G. W., N. Rodenhouse, and P. J. Bohlen. 1990. Role of sustainable agriculture in rural landscapes. In *Sustainable agricultural systems*, ed. C. A. Edwards, R. Lal, P. Madden, R. H. Miller, and G. House, 624-36. Ankeny, IA: Soil and Water Conservation Society.

Barrett, G. W. 1990. Nature's model. *Earthwatch* 9:24-25.

Barrett, G. W., and P. J. Bohlen. 1991. Landscape ecology: Application to conservation of biological diversity. In *Landscape linkages and biodiversity*, ed. W. E. Hudson, in press. Washington, D.C.: Island Press.

Callahan, J. T. 1984. Long-term ecological research. *Bioscience* 34:363-67.

Crosson, P. R., and N. J. Rosenberg. 1989. Strategies for agriculture. *Sci. Am.* 261:128-35.

Deland, M. R. 1991. An ounce of prevention . . . after 20 years of cure. *Environ. Sci. Tech.* 25:561-63.

Evans, F. C. 1956. Ecosystem as the basic unit in ecology. *Science* 123:1127-28.

Fahrig, L., and G. Merriam. 1985. Habitat connectivity and population survival. *Ecology* 66:1762-68.

Franklin, J. F., C. S. Bledsoe, and J. T. Callahan. 1990. Contributions of the Long-Term Ecological Research program. *Bioscience* 40:509-23.

Gale, R. P., and S. M. Cordray. 1991. What should forests sustain? Eight answers. *J. For.* 89:31-36.

Gould, F., and R. E. Stinner. 1984. Insects in heterogeneous habitats. In *Ecological entomology*, ed. C. B. Huffaker and R. L. Rabb, 427-49. New York: John Wiley & Sons.

Henderson, M. T., G. Merriam, and J. F. Wegner. 1985. Patchy environments and species survival: Chipmunks in an agricultural mosaic. *Biol. Conserv* 31:95-105.

Holden, C. 1991. Career trends for the '90s. *Science* 252:1110-17.

Horwith, B. 1985. A role for intercropping in modern agriculture. *Bioscience* 35:286-91.

Institute of Ecology. 1982. *Content and priorities for applied ecological science in the 1980's*. Summary of a workshop sponsored by the National Science Foundation, April 12-13, 1982, Indianapolis, Indiana. Athens, GA: University of Georgia Press.

Jantsch, E. 1972. *Technological planning and social futures*. London: Cassell Associated Business Programmers.

Johnson, P. L. 1977. *An ecosystem paradigm for ecology*. Oak Ridge, TN: Oak Ridge Associated Universities.

Kareiva, P. M. 1982. Experimental and mathematical analysis of herbivore movement: Quantifying the influence of plant spacing and quality on foraging discrimination. *Ecol. Monogr.* 52:261-82.

Kareiva, P. M. 1983. Influence of vegetation texture on herbivore populations: Resource concentration and herbivore movement. In *Variable plants and herbivores in natural and managed systems*, ed. R. F. Denno and M. S. McClure, 259-89. New York: Academic Press.

Kareiva, P. M. 1986. Trivial movement and foraging by crop colonizers. In *Ecological theory and integrated pest management practice*, ed. M. Kogan, 59-82. John Wiley & Sons.

Kemp, J. C., and G. W. Barrett. 1989. Spatial patterning: Impact of uncultivated corridors on arthropod populations within soybean agroecosystems. *Ecology* 70:114-28.

Knight, D. H., and L. L. Wallace. 1989. The Yellowstone fires: Issues in landscape ecology. *Bioscience* 39:700-706.

Levine, M. B., A. T. Hall, G. W. Barrett, and D. H. Taylor. 1989. Heavy metal concentrations during ten years of sludge treatment to an old-field community. *J. Environ. Qual.* 18:411-18.

Lorenz, G. C., and G. W. Barrett. 1990. Influence of simulated corridors on house mouse (*Mus musculus*) dispersal. *Am. Midl. Nat.* 123:348-56.

Lowrance, R., P. F. Hendrix, and E. P. Odum. 1986. A hierarchical approach to sustainable agriculture. *Am. J. Altern. Agric.* 1:169-73.

Lubchenco, J., A. M. Olson, L. B. Brubaker, S. R. Carpenter, M. M. Holland, S. P. Hubbell, S. A. Levin, J. A. MacMahon, P. A. Matson, J. M. Melillo, H. A. Mooney, C. H. Peterson, H. R. Pulliam, L. A. Real, P. J. Regal, and P. G. Risser. 1991. The Sustainable Biosphere Initiative: An ecological research agenda. *Ecology* 72:371-412.

MacMahon, J. A., D. L. Phillips, J. V. Robinson, and D. J. Schimpf. 1978. Levels of biological organization: An organism-centered approach. *Bioscience* 28:700-704.

McElroy, W. D. 1977. The global age: Roles of basic and applied research. *Science* 196:267-70.

Messer, J. J., R. A. Linthurst, and W. S. Overton. 1991. An EPA program for monitoring ecological status and trends. *Environ. Monit. Assess.* 17:67-78.

Myers, P. C. 1988. Conservation at the crossroads. *J. Soil Water Conserv.* 43:10-13.

National Research Council. 1989. *Alternative agriculture*. Washington, D.C.: National Academy Press.

Naveh, Z., and A. S. Lieberman. 1984. *Landscape ecology: Theory and application*. New York: Springer-Verlag.

Odum, E. P. 1971. *Fundamentals of ecology*. Philadelphia: W. B. Saunders Co.

Odum, E. P. 1989. Input management of production systems. *Science* 243:177-82.

O'Neill, R. V., D. L. DeAngelis, J. B. Wade, and T. F. H. Allen. 1986. *A hierarchical concept of ecosystems*. Princeton, NJ: Princeton University Press.

Pickett, S. T. A., and P. S. White. 1985. The ecology of natural disturbances and patch dynamics. San Diego: Academic Press.

Pimentel, D., L. McLaughlin, A. Zepp, B. Lakitan, T. Kraus, P. Kleinman, F. Vancini, W. J. Roach, E. Graap, W. S. Keeton, and G. Selig. 1991. Environmental and economic effects of reducing pesticide use. *Bioscience* 41:402-409.

Power, J. F. 1987. Legumes: Their potential role in agricultural production. *Am. J. Altern. Agric.* 2:69-73.

Risser, P. G., J. R. Karr, and R. T. T. Forman. 1984. *Landscape ecology: Directions and approaches*. Champaign, IL: Illinois Natural History Survey Special Publication, no. 2.

Rowe, J. S. 1961. The level-of-integration concept and ecology. *Ecology* 42:420-27.

Shoemaker, C. A. 1981. Applications of dynamic programming and other optimization methods in pest management. *Inst. Electrical and Electronics Engineers Trans. on Automatic Control* 26:1125-32.

Slobodkin, L. B. 1988. Intellectual problems of applied ecology. *Bioscience* 38: 337-42.

Steiner, F. 1987. Soil conservation policy in the United States. *Environ. Manage.* 11:209-23.

Steinhart, J. S., and C. E. Steinhart. 1974. Energy use in the U.S. food system. *Science* 184:307-16.

Turchin, P. 1986. Modelling the effect of host patch size on Mexican bean beetle *(Epilachna varivestis)* emigration. *Ecology* 67:124-32.

Urban, D. L., R. V. O'Neill, H. H. Shugart, Jr. 1987. Landscape ecology. *Bioscience* 37:119-27.

Wegner, J. F., and G. Merriam. 1979. Movements by birds and small mammals between a wood and adjoining farmland habitats. *J. Appl. Ecol.* 16:349-57.

World Commission on Environment and Development (WCED). 1987. *Our common future*. New York: WCED.

Sustainable Agriculture Research at the Watershed Scale

Richard Lowrance

SUMMARY. Ecological sustainability, the ability of life-support systems to maintain the quality of the environment, is a necessary condition for longterm agricultural sustainability at the field, farm, or national level. Research is needed at the watershed/landscape level to address the effects of changes in inputs, land use, or management practices on the ecological sustainability of modern agriculture. This research should address the impact of perturbations on the responses of watersheds relative to objective indicators of sustainability. Ideally this research would be done within a set of managed landscapes that could be perturbed to change the watershed or landscape response. The perturbations could be to the inputs of materials, energy, and management; the structure of the landscape; or the desired production outputs. An existing set of watershed improvement projects, funded by Federal, State, local, and private sources, could provide the landscapes for this research. Projects funded through Clean Water Act Section 319 grants, USDA Water Quality Demonstration projects, and USDA Hydrologic Unit Area projects provide landscapes throughout the United States that are valuable resources for research on ecological sustainability. These landscapes should be used in research to determine ways to enhance the longterm sustainability of agriculture.

SUSTAINABILITY IN AN ENVIRONMENTAL QUALITY CONTEXT

Agricultural sustainability has been defined in myriad ways. Each author or program administrator seems to provide his or her own answer to the ques-

Richard Lowrance is Ecologist with the U.S. Department of Agriculture-Agricultural Research Service, Tifton, GA 31793.

tion, "What is sustainable agriculture?" The resulting confusion is due to a lack of recognition that sustainability is dependent on the "spatial perspective" of the observer and is a shifting target that will change through time.

We have tried to clarify the questions of "spatial perspective" by defining sustainability within an essentially spatial hierarchy (Lowrance et al. 1986). Within this hierarchy, we have defined the various sustainability objectives that are necessary within modern agricultural systems: agronomic, microeconomic, ecological, and macroeconomic sustainability. Agronomic sustainability is the ability of a tract of land or field to maintain productivity over a long period. Microeconomic sustainability is the ability of a farm enterprise to continue functioning as the basic economic unit. Ecological sustainability is the ability of life-support systems to maintain the quality of the environment and the ability of nonagricultural ecosystems to maintain their ecological integrity. Macroeconomic sustainability is the ability of national production systems to compete in both domestic and foreign markets.

Obviously, research can contribute to achieving sustainable systems at all these levels. As in any hierarchical system, a perturbation at one level may be either a stabilizing or destabilizing force at another level (Allen and Starr 1982). Much of what is recognized as "sustainable agriculture" research is focused on agronomic sustainability. A number of factors contribute to this situation, including the traditional orientation of agricultural science to field-scale research and the difficulty and expense of studying large-scale systems. Conversely, much of the ecological research done by the U.S. Environmental Protection Agency (EPA) has been focused on the behavior of larger-scale systems, and many of the research approaches used for these studies should be adaptable to questions of ecological sustainability.

Society has established some "objective indicators" of unacceptable environmental contamination by establishing drinking water standards, maximum contaminant levels (MCL), total maximum daily load (TMDL) (U.S. EPA 1991), etc. Unfortunately, many of these standards are based on projected human health effects that have little to do with effects of agricultural practices on agroecosystems or adjacent ecosystems. Objective indicators of agricultural sustainability at any of the hierarchical levels are ill-defined. EPA's Environmental Monitoring and Assessment Program (EMAP) is seeking to develop objective indicators of the condition of ecological systems (Messer et al. 1991; Neher, this volume). Indicators identified in EMAP that would be useful indicators of ecological sustainability include: (1) watershed nutrient budgets; (2) soil erosion and deposition rates; (3) land use/extent of noncrop vegetation; (4) cumulative crop yield; and (5) relative abundance, distribution, and demographics of animal populations. Objective indicators such as MCL and TMDL could contribute to development of indicators of ecological sustainability.

Even as we develop objective indicators of sustainability, it is worth noting that society, through the actions of our elected representatives, is likely to make sustainability a shifting target. As scientific understanding of the effects of management practices on adjacent ecosystems and agroecosystems increases, we are likely to see stricter standards that will alter the operational definition of ecological sustainability.

THE WATERSHED AS AN APPROPRIATE SPATIAL SCALE FOR ASSESSING ECOLOGICAL SUSTAINABILITY

Landscape ecology is the study of spatial heterogeneity and spatial and temporal interactions and exchanges across heterogeneous landscapes (Risser et al. 1984). Landscape ecology provides a conceptual framework for addressing the interactions of agricultural and nonagricultural ecosystems that produce landscape- and regional-scale phenomena (Lowrance and Groffman 1987). The pattern of land use and the interactions between components of the landscape produce a landscape response that is the fundamental functional attribute of the landscape. At least three general types of landscape response are possible: (1) response as production variables; (2) response as the movement of organisms and propagules; and (3) response as an environmental quality variable.

The watershed is a special type of landscape that can be defined as providing either the hydrologic input to surface water (surface watershed) or the hydrologic input to ground water (subsurface watershed). The surface and subsurface watersheds may coincide, but most often do not. The watershed integrates land use, cropping systems, climate, soils, hydrogeology, and cultural factors into a "watershed response." The watershed response is essentially a movement of water and water-borne substances. Measurement of watershed response is simpler if there is no movement of water to ground water that is not captured by surface flow on the defined surface watershed. If there is significant recharge to regional ground water, the measurement of the actual watershed response is still conceptually simple but is more difficult in practice. This is why most gaged watersheds are in areas where there is very little recharge to regional ground water.

Watersheds can also be used as the geographic basis for other measurements of landscape structure, function, and response. Inputs have a major role in determining the structure, function, and response of agricultural landscapes. Measurements of these attributes or responses of watersheds must be integrated by the observer rather than depending on hydrologic processes to produce a measurable integrated response. For instance, total net primary production, nutrient input/output ratios, or total density of insect pests can all be measured at the watershed landscape scale, but are not integrated by the process of water movement within the watershed.

IDEALIZED APPROACH TO THE STUDY OF WATERSHED-SCALE SUSTAINABILITY

Given the assumption that objective standards or indicators of ecological sustainability are possible, what are the ways in which we could study sustainability questions at the watershed- or landscape-scale? An idealized approach would be to (1) develop a set of objective watershed- or landscape-scale sustainability indicators; (2) postulate a set of perturbations that would be expected to have an effect on those sustainability indicators; (3) examine the longterm response (as measured by the indicators) of a range of agricultural landscapes to these perturbations; and (4) use these data to determine landscape configurations that enhance ecological sustainability. Ideally, landscapes can eventually be modeled by the use of optimization techniques similar to procedures used in the analysis of large-scale water resource systems, which are actually engineering approaches to controlling landscape structure and function (Haimes 1977; Lowrance and Groffman 1987).

The perturbations to the system could be to either the input or output environment of the system. The input environment provides the materials, land, and information used in the production process. Changes in the input environment will affect the ability of the farmer to obtain and use inputs such as seed, fertilizers, agrichemicals, machinery, land, water, management, and financing. The output environment includes the market and price supports that control the prices of commodities as well as the regulations/public perceptions that control or influence the production of externalities such as water pollution or chemical wastes and chemical residues.

Input perturbations could be at any of the hierarchical levels discussed above. Examples of these perturbations abound because they are the types of changes to which modern agriculture is constantly forced to adjust. Examples of input perturbations would be: (1) a large increase in the price of nitrogen fertilizer or pesticides or a loss of pesticides due to loss of registration (agronomic scale); (2) restrictions on the amount of irrigation water or chemical inputs used per farm (microeconomic scale); (3) restrictions or regulations on land use that affect the use of land for agricultural purposes (ecological scale); or (4) a rise in the cost of petroleum products due to changes in world oil supply (macroeconomic scale). Examples of output perturbations would generally be related to either changes in the price of crops (generally controlled by either macroeconomic or governmental factors) or a change in the allowable outputs of byproducts to air, water, or soil. From an agricultural perspective, changes in the output environment are, by definition, at the ecological or macroeconomic level.

There are obviously many practical difficulties to this idealized approach. First of all, one would need a large number of experimental landscapes to represent the different important agricultural regions of the United States. Secondly, the perturbations will change through time. Some changes will be permanent (e.g., loss of a registered pesticide) while some perturbations will be transient (e.g., changes in crop prices). In addition, developing technologies can affect the level and type of inputs and the efficiency of output/input relationships. Finally, a substantial commitment of resources would be needed to identify, manipulate, and measure landscape or watershed responses.

TWO PRACTICAL APPROACHES TO THE STUDY OF ECOLOGICAL SUSTAINABILITY AT THE WATERSHED OR LANDSCAPE LEVEL

Given the difficulties of the idealized approach, what practical alternatives exist for developing information on ecological sustainability? At least two practical approaches are possible: (1) a statistical approach where one analyzes a range of existing landscapes whose input/output environments serve as surrogates for the perturbations of interest; (2) a design or restoration approach where selected watersheds that are "less sustainable" are modified to enhance the performance of sustainability indicators. Both of these approaches will require data collection on the watershed response, landscape structure, distribution of cropping systems, etc.

The statistical approach would require selection of multiple landscapes in similar hydrologic settings that are likely to reflect a range of ecological sustainability. Differences in watershed responses and sustainability indicators would be determined. These differences would be statistically related to differences in such factors as land use distribution, extent and distribution of buffer systems, temporal patterns of crops and inputs, and other factors that control the watershed response. One major difficulty with this method would be finding a wide range of agricultural practices and landscape structures within a given ecoregion and geohydrologic setting.

The design/restoration approach would use a more limited set of landscapes but would provide information on changes in sustainability indicators and watershed responses following changes in field and watershed scale practices. The restoration/design approach would require gathering of similar data on landscape structure and function, cropping patterns, input management, etc., but would attempt to take a before-and-after approach on watersheds that were not ecologically sustainable based on objective indicators and the watershed response. Management practices to enhance ecological sustainability would be applied in these landscapes. Effects on agronomic, microeconomic, and ecological sustainability could be determined simultaneously if appropriate field-, farm-, and landscape-scale measurements were made.

One possibility for implementing the design/restoration approach is to incorporate research on ecological sustainability into existing "Watershed Improvement Projects" funded by EPA and USDA. There are at least three types of these projects active today. Through money passed to the states under Section 319 of the Clean Water Act, EPA funds nonpoint source pollution control projects. Section 319 funds are used to provide cost-share incentives for Best Management Practices (BMPs), technical assistance, and limited monitoring for a wide range of watershed improvement projects. These projects are not designed to have a research component, but many of the land-use and management practice changes needed for restoration and enhancement of ecological sustainability can or could be applied through Section 319 projects. Successful Section 319 projects usually involve extensive coordination among State natural resource agencies, USDA, soil and water conservation districts, and other private and public groups.

USDA funds Water Quality Demonstration projects and Hydrologic Unit Area projects (HUAs), which target specific watersheds with water quality problems (Mussman 1990). There are presently 16 USDA Water Quality Demonstration projects funded. The Extension Service is generally the lead agency on these projects. These projects are designed to provide technical assistance and educational resources to farmers/land managers in the project watersheds. It is also typical for information to be gathered on attitudes toward water quality/environmental quality problems, so that the success of educational and outreach efforts can be judged. Hydrologic Unit Area projects are more numerous than the Water Quality Demonstration projects and are designed to address a specific set of water quality concerns or a specific water quality problem. The Soil Conservation Service (SCS) is often the lead agency on HUAs. Cost-share funding can be provided for BMPs which are SCS-approved practices. There can be considerable overlap between USDA projects (demonstrations or HUAs) and Section 319 water quality projects. Some demonstration projects are also Section 319 projects and Section 319 funds are used for implementing specific practices with HUAs.

CONCLUSIONS

Although water quality improvement projects address just one aspect of ecological sustainability, it is an aspect of considerable public concern. It would be possible to address ecological sustainability without including water quality parameters as objective indicators, but the omission would reduce the relevance of the research to society's concerns. Incorporating new approaches into water quality improvement projects to achieve changes in other sustainability indicators offers a cost-effective means of conducting research on ecological sustainability in agricultural watersheds. Agencies such as EPA with interests in large-scale ecological research could leverage their research funds by designing studies that build upon existing water quality improvement projects.

REFERENCES

Allen, T. F. H., and T. B. Starr. 1982. *Hierarchy: Perspectives for ecological complexity.* Chicago: University of Chicago Press.

Haimes, Y. Y. 1977. *Hierarchical analyses of water resources systems.* New York: McGraw-Hill.

Lowrance, R., and P. M. Groffman. 1987. Impacts of low and high input agriculture on landscape structure and function. *Am. J. Altern. Agric.* 2:175-83.

Lowrance, R., P. F. Hendrix, and E. P. Odum. 1986. A hierarchical approach to sustainable agriculture. *Am. J. Altern. Agric.* 1:169-73.

Messer, J. J., R. A. Linthurst, and W. S. Overton. 1991. An EPA program for monitoring ecological status and trends. *Environ. Monit. Assess.* 17:67-78.

Mussman, H. C. 1990. Solutions to water quality problems: USDA's role. *J. Soil Water Conserv.* 45(6):598-99.

Risser, P. G., J. R. Karr, and R. T. T. Forman. 1984. *Landscape ecology: Directions and approaches.* Champaign, IL: Illinois Natural History Survey, Special Publication, no. 2.

U.S. EPA. 1991. Some observations and a report on the newly issued guidance for water quality-based decisions: The TMDL process. *Nonpoint Source News-Notes* 12:4-6.

Addressing Information Needs to Support Sustainable Agriculture Policies

Clayton W. Ogg

SUMMARY. Policy makers are more willing to adopt changes when the benefits of the change are clearly documented. The 1977 National Resources Inventory (NRI) and the dramatic changes it helped bring about is an example of the power of information in facilitating environmental policy innovation for agriculture. Multibillion-dollar conservation programs now address soil erosion in the United States in a relatively strategic and technically sophisticated way. This greater integration between farm policy and environmental policy has had an impact on commodity policies for agriculture, which have been modified in an attempt to avoid barriers to sustainable agriculture crop rotations. However, identifying the benefits of sustainable farming systems presents a greater research challenge than the soil conservation studies that used the NRI. The rewards for policy-oriented ecological research are great because many opportunities remain for U.S. and world agriculture to become more supportive of environmental and ecological goals while simultaneously meeting farmer and consumer needs.

INTRODUCTION

Environmental policy innovation for agriculture requires good ideas but also information to anticipate impacts on agriculture and the environment. Scientific information can play a decisive role, as exemplified by the dramatic changes in conservation, commodity, and environmental policy that were precipitated by the explosion of information resulting from the 1977 National Resources Inventory (NRI) (USDA 1980).

Clayton W. Ogg is a staff member in the Office of Policy Analysis, U.S. Environmental Protection Agency, 401 M Street, SW, P.M. 221, Washington, DC 20460.

This information allowed policy makers to identify the need for and benefits from a multibillion-dollar Conservation Reserve Program (CRP). The CRP now idles one-tenth of U.S. croplands, drawing from the highly erodible lands that account for most of the soil erosion in the United States. The early CRP proposals' popularity provided momentum for a number of related initiatives directed at highly erodible lands and wetlands, including compliance programs, and more recently, linkages between federally funded soil conservation programs and state-led water quality programs (Dicks et al. 1988; Ogg 1988).

Analysts and policy makers next considered policies to support adoption of sustainable agriculture technologies. Sustainable agriculture technologies contribute to a much broader set of goals than the CRP and compliance programs, including soil productivity, water quality, farmer and consumer safety, and ecological as well as cost-saving goals. Once again, information played a critical role, anticipating at least some of the likely impacts on producers and the environment of the new flexibility (Hertel et al. 1990; Ogg 1990; Young and Goldstein 1989) offered to farmers, to a limited degree, in the 1990 Farm Bill's commodity provisions.

Multiple objectives add to the challenge of predicting impacts. New Integrated Farm Management and Water Quality Incentives programs remain largely unfunded. Research documenting potential contributions to farmers, consumers, and ecological systems could facilitate effective implementation of these innovative programs.

INFORMATION AND ANALYSIS LEADING UP TO THE FOOD SECURITY ACT OF 1985

Policy makers now rely on analysis and pilot programs to test ideas that might otherwise seem complex or risky to implement nationally. The most striking example of new information and analysis leading to innovative new environmental programs was provided by the CRP-related analyses using the 1977 NRI.

The 1977 NRI not only provided the first objective picture of soil erosion problems in the United States, but it also furnished the underlying parameters, such as erodibility, slope, and cover practices on each site sampled, to anticipate the effects of program changes. Targeting the more serious problems became a major issue (USDA 1981a, 1981b), as did the lack of conservation program tools to treat the land that accounted for the worst erosion problems. We learned, for example, that erosion problems are concentrated on a small portion of the Nation's cropland, which is, however, dispersed across the country in ways that were expeditious for implementing a CRP-type program (Schaller et al. 1985; Ogg et al. 1982).

Past economic studies of an earlier conservation reserve (Bottum et al. 1961, Brandow 1961, Brandow 1977, Robinson 1966) documented large economic advantages of a conservation reserve structure for idling land versus the commodity programs' structure. By the early 1980s, these commodity program costs were reaching an unprecedented 20 to 30 billion dollars per year. Research that incorporated the new NRI data into a mathematical programing model (Shaller et al. 1985; Ogg et al. 1984) and related research (Webb et al. 1985) discovered that much of the same cost savings suggested by the studies cited above were possible from the CRP if it was targeted to highly erodible land. Commodity programs, such as the Payment in Kind program, could idle only a small portion of each participating farm and paid every farmer what it cost to attract the most reluctant participant.

Sodbuster, Swampbuster, and Conservation Compliance programs complemented the CRP and commodity program goals by discouraging misuse of fragile lands, whose use in crop production undermined commodity programs' attempts to reduce crop surpluses (Ogg 1986). Enthusiasm for what the CRP was expected to accomplish had an impact on these complimentary proposals.

With all of the interest in saving billions of dollars and eliminating millions of tons of soil erosion, policy makers in the early 1980s appeared less alert to studies of potential ecological effects. Yet, one study documented the advantages of a CRP-type program to reduce eutrophication problems in a water supply reservoir (Ogg et al. 1983). Idling some of the most erodible land, at least part of the time, is a critical ingredient of efforts to address eutrophication problems where substantial reductions in phosphorus loads are needed.

In the years leading up to the 1990 Farm Bill, policy makers' interests clearly changed to concentrate increasingly on ecological goals, as evidenced by a new riparian filter, wetland, and wildlife focus for the CRP and by sustainable agriculture programs that focus on ecological goals. Providing information to support these multi-objective programs represents a far greater research challenge than providing information on the conservation programs.

PROVIDING INFORMATION TO SUPPORT MULTIPLE-OBJECTIVE PROGRAMS

Discretionary provisions allowing the Secretary of Agriculture to designate CRP eligibility to lands that contribute to other environmental problems besides soil erosion were used to a limited degree before 1990. Enrollment of cropped wetlands occurred, and vegetative filters along streams also became eligible, although few acres of filters were enrolled.

However, the 1990 Farm Bill made water quality the top priority, replacing soil conservation, since the final 6 million acres of the 40 million acre CRP will address multiple wildlife and ecological objectives. This program change and the largely unfunded sustainable agriculture programs highlight the need for information regarding water quality and ecological benefits.

CRP INFORMATION GAPS

Some research suggests tremendous ecological benefits from the relatively small acreage that is needed to provide vegetative filters along streams (Lowrance et al. 1984; Schipper et al. 1990). Filters can be most effective by filtering sediment, denitrifying water from seeps, and providing a strategically placed wildlife corridor along streams. While these findings would very strongly favor focusing the remaining CRP acreage along streams, hardly any enrollment of filters or riparian corridors has occurred to date. Further documentation of ecological benefits would help attain the needed focus during the implementation of these programs.

Research is also needed regarding filter design. Dillaha et al. (1989) point out that much of the water entering streams arrives as a concentrated flow that will plow through filters along streams. The current CRP addresses this problem imperfectly by making grassed waterways eligible to begin to filter water entering the concentrated flow areas. Dillaha et al. (1989) also raise concerns about sediment build-up along those streams bordering the most erosive fields. Riparian land use rather than mowed grass may provide many of the filtering benefits of grass, while emphasizing benefits to wildlife. The U.S. Environmental Protection Agency (EPA) and the U.S. Department of Agriculture (USDA) jointly pursue these filter design questions through cooperative research projects.

SUSTAINABLE AGRICULTURE INFORMATION GAPS

Uncertainty represents a relatively serious constraint to effectively implementing sustainable agriculture programs. Definitional problems are adding to uncertainties, which affect funding.

Broad agreement exists that crop rotations dramatically reduce the need for insecticides (CARD 1988) and, in some instances, reduce the need for commercial fertilizer and herbicides. Nitrogen is a pervasive pollutant of groundwater within intensively farmed areas (Madison and Brunett 1985), and herbicides in surface water often exceed safe drinking levels throughout major midwestern rivers (Goolesby et al. 1989, Richards and Baker 1989).

Timing fertilizer applications to correspond to plant needs during the growing season (Bouldin et al. 1971) and soil testing (Fox and Piekielek 1983; Magdoff et al. 1984) help reduce nitrate leaching, although much research and calibration is needed to improve nitrogen fertilizer recommendations in order to realize these water quality and economic benefits. Choice of tillage methods, including ridge till, and placement can facilitate reduced herbicide use and reduce runoff (Kramer et al. 1989). Using less chemically intensive practices contributes to pollution prevention and to attaining the environmental goals of sustainable agriculture.

Aspects of farm programs that hindered farmers in practicing crop rotations and reducing chemical and other input use were identified and cited during the 1990 Farm Bill debates; studies also identified some of the beneficial impacts on farm programs of removing barriers to lower input use (Hertel et al. 1990; Ogg 1990; Young and Goldstein 1989). Although one or two commodity groups apparently remain concerned that the outcome for their commodity is not understood with complete certainty, research shows that past farm programs undermined their own price support objectives by encouraging input use (Hertel et al. 1990).

Considerable flexibility was ultimately built into the law, particularly during the budget negotiations. Thirty percent of the base acreage used in price support payment formulas can now be planted to another crop, without loss of base. Yields from past years, used in calculating farmers' payments (payment yields), also remain frozen at the 1985 farm program levels. Farmers, therefore, have no assurance that applying more chemicals to raise yields will increase government payments.

These reforms were accomplished in spite of imperfect knowledge of the economic and ecological impacts. As potential ecological and health benefits of sustainable agriculture crop rotations and lower input practices are further documented, they will be increasingly relevant to the ongoing discussions concerning commodity program reform.

Essentially the same information needs arise as ways to support adoption of integrated crop management (ICM) systems are considered. These sustainable farming practices appear under various names in the 1990 Farm Bill but have received very limited funding. The ICM pilot, conducted by the Agricultural Stabilization and Conservation Service, cost shares private consultants' advice on reducing input use. As policy makers consider expanding this promising program, research might contribute by identifying the value of such information to farmers and the environment. ICM consultants essentially earn their living by reducing farmers' expenditures on inputs that are associated with environmental risks, such as fertilizer and pesticides. The Extension Service provides training programs aimed at various consultants and has worked with them.

Quantifying the variety of ecological gains from sustainable practices presents a lengthy list of research challenges. As a starting point, more information is needed about chemical input savings and yield effects of adopting crop rotations and other ecologically sound farming practices. Such savings may result in considerably larger reductions in leaching relative to the reduction in fertilizer use (Davis and Heatwole 1990). Therefore, a greater research focus on the ultimate water quality benefits appears promising.

Where and how widely are the more ecologically sound practices applied and what are the trends in adoption? Such information needs to be developed by region, as sustainable practices and their need vary by region. In fact, USDA is currently expanding efforts to obtain information about practices.

More specific information gaps include the ecological trade-offs between reduced tillage, which is favored by farm plans required by Conservation Compliance, and ridge-tillage cultivation. The latter practice has received far less attention and support from conservation policy. Yet, ridge-tillage has the potential to accommodate much less reliance on herbicides, as well as reducing soil erosion (at least when applied on the contour) (Kramer et al. 1989).

An issue addressed by an ongoing National Academy of Science (NAS) study concerns the improved soil productivity that results from many sustainable practices and the effects of this enhanced productivity on water quality and the environment. Many sustainable practices improve water quality and soil productivity. Whether this enhanced productivity is a source of further environmental gains is the hypothesis that needs testing. Healthier soil, rich in organic matter, may allow plants to better utilize nutrients that might otherwise leach. (Or, under the alternative hypothesis, leaching will increase due to increased infiltration). Healthier soil is also hypothesized to provide global climate benefits by retaining more carbon. Results of this congressionally mandated NAS study will be presented to EPA in a report this year and at regional conferences.

Early results from cooperative work on consumer preferences for sustainable agriculture products also suggest that government subsidies and soil productivity gains may not be the only potential incentive to adopt such practices. Consumers would be willing to pay substantially more for products produced without pesticide spraying or partially free of pesticides, provided high cosmetic quality could also be achieved (Van Ravenswaay and Hoehn 1990). To make such produce available requires overcoming significant challenges in market organization as well as production, but the rewards are also great. Consumer demands could provide a market incentive for farmers to adopt practices that meet some of the more ambitious sustainable agriculture goals. Information about consumer willingness to pay needs further documentation, as does analysis of actual health risks posed by low levels of pesticide residues on food.

CONCLUSIONS

Policy innovation requires providing information to policy makers that anticipates the major consequences of implementing policy changes. For policy options supporting sustainability, interrelated economic, ecological, soil productivity, and global climate effects need analysis. Challenges of providing this information in a timely way are considerable due to the multiple objectives addressed by sustainable agriculture programs. We can increase our effectiveness in providing this information by increasing communication and joint research across disciplines, conducting cooperative research efforts, and being alert to institutional changes that occur at all levels of government.

REFERENCES

Bottum, J. C., J. O. Dunbar, R. L. Kohls, D. L. Vogelsang, G. McMurtry, and S. E. Mogan. 1961. *Land retirement and farm policy*. Agricultural Experiment Station Bulletin 704. West Lafayette, IN: Purdue University.

Bouldin, D., W. Reid, and D. Lathwell. 1971. Fertilizer practices which minimize nutrient loss. Agricultural wastes: Principles and guidelines for practical solutions. In *Proceedings of Cornell University conference on agricultural waste management* at Syracuse, New York, 192-208. Ithaca, New York: Cornell University Department of Engineering.

Brandow, G. 1961. Reshaping farm policy. *J. Farm Econ.* 43:1019-31.

Brandow, G. 1977. Policy for commercial agriculture, 1945-71. In *A survey of agricultural economics literature*, vol. I, ed. L. E. Martin, 207-92. Minneapolis: University of Minnesota Press.

Center for Agriculture and Rural Development. 1988. *Corn root worm analysis: An application of CEEPS*. Ames, IA: Iowa State University.

Davis, P. E., and C. D. Heatwole. 1990. Modeling impacts of alternative agricultural systems on groundwater in the Virginia coastal plain. Presented at the International Winter Meetings of the American Society of Agricultural Engineers, December 18-21, Chicago, Illinois. Paper no. 902592.

Dicks, M. R., F. Llacuna, and M. Linsenbigler. 1988. *The conservation reserve program: Implementation and accomplishments.* USDA Statistical Bulletin no. 763. Washington, D.C.: U.S. Department of Agriculture, Economic Research Service.

Dillaha, T. A., J. H. Sherrard, and D. Lee. 1989. Long term effectiveness of vegetative filter strips. *Water, Environment, and Technology*, November: 418-21.

Fox, R. H., and W. P. Piekielek. 1983. *Response of corn to nitrogen fertilizer and the prediction of soil nitrogen availability with chemical tests in Pennsylvania.* Pennsylvania Agricultural Experiment Station Bulletin 843. University Park, PA: Pennsylvania State University.

Goolesby, D. A., E. M. Thurman, D. W. Kolpin, and M. G. Betnoy. 1989. A reconnaissance for trizene herbicides in surface waters of the Upper Midwest, U.S. In *Proceedings of the 46th annual meeting of the Upper Mississippi River Conservation Committee.* Washington, D.C.: U.S. Geological Survey.

Hertel, T. W., M. E. Tsigas, and P. V. Preckel. 1990. Unfreezing program payment yields: Consequences and alternatives. *Choices*, Second Quarter: 32-33.

Kramer, L. A., A. T. Hjelmfelt, Jr., and E. E. Alberts. 1989. *Watershed runoff and nitrogen loss from ridge-till and conventional till corn.* American Society of Agricultural Engineers, St. Joseph, MO. Paper no. 89-2502.

Lowrance, R., R. Todd, J. Fail, Jr., O. Hendrickson, Jr., R. Leonard, and L. Asmussen. 1984. Riparian forests as nutrient filters in agricultural watersheds. *Bioscience* 34:374-77.

Madison, R. J., and J. O. Brunett. 1985. Overview of the occurrence of nitrate in ground water of the United States. In *USGS National Water Summary*, 93-105. USGS Water Supply Paper 2275. Washington, D.C.: U.S. Geological Survey.

Magdoff, F. R., D. Ross, and J. Amadon. 1984. A soil test for nitrogen availability to corn. *Soil Sci. Soc. Am. J.* 48:1301-1304.

Ogg, C. W. 1986. New cropland in the NRI: Implications for resource policy. In *Soil conservation: Assessing the National Resources Inventory*, vol. 2, 253-69. Washington, D.C.: National Academy Press.

Ogg, C. W. 1988. *The Conservation Title of the Food Security Act of 1985: Challenge of a multiple-objective program.* USDA Staff Paper No. AGES880413. Washington, D.C.: U.S. Department of Agriculture.

Ogg, C. W. 1990. Farm price distortions, chemical use, and the environment. *J. Soil Water Conserv.* 45:45-47.

Ogg, C. W., J. D. Johnson, and K. C. Clayton. 1982. Policy options for targeting conservation expenditures to the most erosive soils. *J. Soil Water Conserv.* 37:68-72.

Ogg, C., H. Pionke, and R. Heimlich. 1983. A linear programming economic analysis of lake quality improvements using phosphorus buffer curves. *Water Resour. Res.* 19:21-31.

Ogg, C. W., S. E. Webb, and W. Y. Huang. 1984. Cropland acreage reduction alternatives: An economic analysis of a soil conservation reserve and competitive bids. *J. Soil Water Conserv.* 39:379-83.

Richards, P., and D. B. Baker. 1989. Potential for reducing human exposures to herbicides by selective treatment of storm runoff water at municipal water supplies. In *Proceedings of a national conference on pesticides in terrestrial and aquatic environments*, ed. D. Weigmann. Blacksburg, VA: Virginia Polytechnic Institute and State University.

Robinson, K. 1966. Cost and effectiveness of recent government land retirement programs in the United States. *J. Farm Econ.* 48:22-30.

Schaller, N., R. Clark, W. Huang, C. Ogg, and S. Webb. 1985. *Analysis of policies to conserve soil and reduce surplus crop production.* USDA Agricultural Economic Report no. 534. Washington, D.C.: U.S. Department of Agriculture, Economic Research Service.

Schipper, L., A. Cooper, and W. Dyck. 1990. Mitigating non-point source nitrate pollution by riparian zone denitrification. Paper presented at the NATO Conference, Lincoln, Nebraska, August, 1990.

U.S. Department of Agriculture. 1981a. *1980 Appraisal. Part I. Soil, water and related resources in the United State–analysis of conditions and trends.* Washington, D.C.: USDA.

U.S. Department of Agriculture. 1981b. *National summary evaluation of the agricultural conservation program, phase 1.* Washington, D.C.: USDA.

U.S. Department of Agriculture. 1980. National resources inventory, computer tapes. Washington, D.C.: USDA, Soil Conservation Service.

Van Ravenswaay, E. O., and J. P. Hoehn. 1990. *The impact of health risk on food demand: A case study of alar and apples.* East Lansing, MI: Department of Agricultural Economics, Michigan State University. Staff Paper 90-31.

Webb, S., C. Ogg, and W. Huang. 1985. *Idling erodible cropland: Impacts on production, prices, and government costs.* Agricultural Economic Report no. 550. Washington, D.C.: U.S. Department of Agriculture, Economic Research Service.

Young, D. L., and W. A. Goldstein. 1989. How government programs discourage sustainable cropping systems: A U.S. case study. Paper presented at Farming Systems Research Symposium, October 8, 1989, Fayetteville, Arkansas.

Reducing Agricultural Impacts on the Environment: Current EPA Program and Research Activities– And Future Directions

Gail M. Robarge
Jay Benforado

SUMMARY. Successful adoption of sustainable agricultural practices in the United States will hinge on identifying the potential benefits of these practices and demonstrating their feasibility within the context of agricultural economic conditions. The U.S. Environmental Protection Agency (EPA) has a number of ongoing research efforts that relate to sustainable agriculture. However, these efforts have not been designed as, and do not constitute, an organized environmental/agricultural research program. New research initiatives, such as the Environmental Monitoring and Assessment Program, are evolving towards a more integrated approach for solving environmental/agricultural problems. These efforts provide a foundation for an integrated program in sustainable agriculture research. One of the major challenges will be to work effectively with the U.S. Department of Agriculture, taking advantage of the research capabilities and strengths of each agency.

INTRODUCTION

Agroecosystems are our most intensively managed natural resource, and a large portion (approximately 400 million hectares) of the United States land

Gail M. Robarge and Jay Benforado are staff members with the Office of Technology Transfer and Regulatory Support, U.S. Environmental Protection Agency, 401 M Street, SW, Washington, DC 20460. The views expressed in this paper are those of the authors and are not intended to represent those of the U.S. Environmental Protection Agency.

area is devoted to agricultural and livestock production. The prevalence of this land use has resulted in a broad spectrum of impacts on human health and the environment including:

- surface water and groundwater contamination by pesticides, fertilizers, and animal waste;
- soil erosion, which reduces soil tilth on cropland and leads to sedimentation in streams, lakes, and estuaries;
- irrigation impacts on water quantity (e.g., reducing in-stream flows and depleting aquifers) and soil and water quality (e.g., salinization of soils, discharge of irrigation return flows to wetlands);
- loss of terrestrial and aquatic habitat due to conversion to cropland and grazing operations, and subsequent reductions in wildlife populations and biodiversity;
- human health impacts due to exposure to agricultural chemicals, including application of chemicals by farm workers, residues of pesticides in food, pesticides and nitrates in drinking water, and home/lawn/garden application of pesticides; and
- links to global climate change: the clearing of land releases stored carbon, and soil carbon levels are affected by land use management, including agricultural practices.

Obviously, agricultural impacts on the environment are significant and diverse. However, U.S. Environmental Protection Agency (EPA) programs, organized to address environmental problems by media (air and water) or by contaminant (pesticides and toxic chemicals), have not comprehensively addressed the environmental impacts of agriculture. Thus, in a 1990 report prepared by EPA's Science Advisory Board (SAB), impacts from agriculture emerged as some of the major unaddressed environmental risks (U.S. EPA 1990a). In its report, the Board presented a comparative ranking of environmental risks (Table 1). Factors affecting the ranking included scope of the problem, length of time required for mitigation, and whether the effects are irreversible.

Three out of the four risks to ecology and human welfare ranked as "relatively high" are related in part to agricultural activities: habitat alteration and destruction, species extinction and overall loss of biological diversity, and global climate change. Also three of the four "medium" risks result, at least in part, from agricultural activities: herbicides and pesticides; toxics, nutrients, biochemical oxygen demand, and turbidity in surface waters; and airborne toxics. Of the four relatively high risks to human health, two are contributed to by agriculture: worker exposure to chemicals, and

TABLE 1. Science Advisory Board Risk Rankings (U.S. EPA 1990a).

Risks to the Natural Ecology and Human Welfare	
Relatively High Risk	Habitat alteration and destruction Species extinction and overall loss of biodiversity Stratospheric ozone depletion Global climate change
Relatively Medium Risk	Herbicides/pesticides Toxics, nutrients, biochemical oxygen demand, and turbidity in surface waters Acid deposition Airborne toxics
Relatively Low Risk	Oil spills Groundwater pollution Radionuclides Acid runoff to surface waters Thermal pollution
Risks to Human Health	
Relatively High Risk	Ambient air pollutants Worker exposure to chemicals in industry and agriculture Pollution indoors Pollutants in drinking water
Potentially significant	Pesticide residues on food and toxic chemicals in consumer products

pollutants in drinking water. Pesticide residues in foods were also identified as potentially significant, although the data in this area were not considered as strong as for the other human health risks.

The SAB recommended that EPA target its environmental protection efforts on the basis of opportunities for the greatest risk reduction; reducing agricultural impacts on the environment offers the potential to reduce many of the risks identified. Other important SAB recommendations pertinent to agriculture included placing greater emphasis on reducing ecological risks, emphasizing pollution prevention, and integrating environmental considerations into policy decision-making in areas such as taxes, energy, agriculture, and international affairs.

The challenge to sustainable agriculture is to address the significant environmental risks without sacrificing the benefit of low cost and abundant food. The potential benefits of sustainable agriculture are manifold; practices that improve soil tilth, for example, also reduce soil erosion and subsequent sedimentation in surface waters, conserve soil moisture, enhance soil fertility, improve agricultural productivity, and potentially help to mitigate global warming by enabling retention of more carbon in the soil. Recognizing and quantifying the environmental benefits that sustainable agriculture may produce is a necessary step in furthering the adoption of these systems.

EPA POLICY PRIORITIES AND PROGRAMS

Current EPA Regulatory Programs

Over the past 20 years, EPA programs have developed in response to the identification of one pollution problem at a time. Congress passed laws to address each problem, and EPA was charged with implementing the specific requirements in the laws. Each EPA program office is responsible for the implementation of one or more laws addressing a particular type of pollution: air, water, waste, and pesticides/toxics. Thus, EPA does not have a comprehensive program that deals with agricultural impacts on the environment but rather has several programs, each of which are assigned a small piece of the agricultural issue by the environmental statutes. The most significant EPA programs related to agriculture are:

- *Regulation of pesticides.* Under the Federal Insecticide, Fungicide and Rodenticide Act (FIFRA), EPA regulates the use of all pesticides, including agricultural and nonagricultural chemicals. In registering pesticides for use, EPA must evaluate both the risks and benefits of pesticide use. Difficulties include evaluating the ecological risks of pesticide use, and identifying alternatives to the pesticide that might be less risky.
- *Biotechnology.* EPA reviews applications for the experimental field release of genetically engineered organisms in order to ensure that appropriate safeguards are taken to protect the environment. EPA is developing regulations for biotechnology products under both the Toxic Substances Control Act, which provides for review of new chemicals before they are manufactured, and FIFRA, under which all pesticidal agents must be registered.
- *Nonpoint source management plans.* Under the Clean Water Act, States are required to assess nonpoint source pollution problems and adopt management programs to control nonpoint source pollution. EPA reviews and approves the State plans and provides grants to assist in implementing the programs. Agriculture has been identified as a major source of nonpoint source pollution, affecting over 50 percent

of river miles that have been identified as not meeting their designated uses, e.g., not fishable or swimmable (U.S. EPA 1989).

- *Wellhead protection program.* The Safe Drinking Water Act requires States to develop wellhead protection programs for public drinking water supply wells. The programs are designed to protect the area around the wellhead from all sources of contamination, including agriculture. EPA provides methods and tools for States to use in delineating wellhead protection areas.
- *Drinking water standards.* EPA establishes enforceable standards for contaminants in public drinking water supplies, including nitrates and some pesticides. In January 1991, maximum contaminant levels (MCLs) were established for 13 pesticides. Health advisory levels (which are not enforceable) have been established for many pesticides, and MCLs for additional pesticides will be established in the future.
- *Wetlands program.* Under Section 404 of the Clean Water Act, EPA has authority to review Army Corps of Engineers permits for dredging and filling of wetlands. Historically, agriculture has caused nearly 90 percent of total wetlands loss, but regulation of wetlands under the Clean Water Act has been controversial.
- *Livestock waste management.* In some regions of the country, animal wastes pose a significant threat to water quality. Under the Clean Water Act, EPA regulates point source discharges to surface waters. Current regulations mandate construction of storage facilities for all wastes and wastewaters from large, concentrated animal feeding operations.

This overview of EPA programs indicates that EPA is not comprehensively addressing the most significant environmental risks posed by agriculture. Some risks are not being adequately addressed–such as pesticide effects on ecosystems, airborne toxics, and pesticide residues in food–and some risks are not being addressed at all–such as terrestrial habitat alteration/destruction and loss of biodiversity. Thus, we cannot use EPA's existing programs as a framework within which to define EPA research needs in sustainable agriculture. We need to move forward from the chemical-specific, medium-specific approach to environmental problems and look at the broader picture.

New EPA Policy Directions

EPA is entering a new phase in environmental protection. In the past, many actions have been taken to address air and water pollution and waste disposal. While often successful, environmental protection policies to solve problems such as these were usually reactive rather than preventive, addressed only one environmental medium at a time, and were not always based on a sound scientific foundation. EPA is now faced with problems, such as those related to agriculture, that are harder to understand and control: nonpoint source water pollution, long-range air transport of pesticides, pesticide residues in food, loss of terrestrial and aquatic biodiversity, climate change, and loss of wetland resources. Steps to address problems such as these are likely to be more successful if EPA does not rely solely on its traditional regulatory approach.

The Agency recently adopted a new "Strategic Direction" statement that describes the EPA's mission and goals for protecting the environment (U.S. EPA 1991a). Guiding principles that are highlighted in the statement include:

- using sound science to develop solutions to environmental problems;
- evaluating health and ecological risks, working to reduce the greatest risks, and measuring the environmental results;
- preventing pollution at the source, rather than controlling pollution after it has been generated;
- protecting the environment as a whole by accounting for cross-media impacts;
- targeting environmental protection efforts in geographic areas with the greatest risks;
- working with other Federal, State, and local government agencies to develop and implement environmentally sound policies; and
- incorporating the costs of environmental impacts into economic decision processes.

The Strategic Direction describes a new way for EPA to achieve its environmental protection goals. Up to this point, EPA has relied almost solely on "command and control" approaches: publishing regulations, writing permits, enforcing the laws. Now, EPA expects to adopt other approaches such as utilizing market mechanisms and economic incentives, developing technological solutions, and providing information to the public and private sectors to encourage environmentally responsible decision-making.

The guiding principles of the Strategic Direction are directly relevant to defining EPA's role in sustainable agriculture research. Consideration of these principles provokes questions such as: What are the environmental science research areas related to agriculture in which EPA needs to excel?

What is EPA's niche as contrasted with the U.S. Department of Agriculture (USDA) research role? How can EPA best work with USDA to ascertain whether new USDA programs to address environmental problems are having a positive impact?

Emerging EPA Programs

Several new EPA programs related to agricultural problems are attempting to implement some of the new principles articulated in the Strategic Direction. The programs highlighted below reflect the new principles in that they are not traditional "command and control" programs, but place greater emphasis on pollution prevention, coordination with other agencies and, to some extent, consideration of cross-media impacts. For example:

- *Pesticides in Ground Water Strategy* (proposed). Proposed in 1988 and likely to be finalized in 1991, this strategy will be used to structure State-level management of pesticides that are likely to leach into groundwater (U.S. EPA 1991b). The strategy focuses on prevention of groundwater contamination, recognizing that cleanup after contamination is difficult and expensive. As States begin to develop and implement pesticide management plans, they will need information on which agricultural practices will best protect groundwater and ecological resources.
- *Guidance for Coastal Nonpoint Source Pollution* (proposed). The Coastal Zone Act Reauthorization Amendments of 1990 require that States with approved coastal zone management programs develop Coastal Nonpoint Pollution Control Programs in order to ensure protection and restoration of coastal waters. The law requires EPA to publish guidance specifying management measures for States to use in addressing nonpoint source water pollution problems in coastal areas. The guidance, published for comment in June 1991, describes management measures in terms of management systems as opposed to focusing on individual practices (U.S. EPA 1991c). The measures focus on pollution prevention activities such as minimizing the application of chemicals and minimizing soil erosion, and include consideration of both surface water and groundwater impacts.
- *Geographic initiatives.* EPA has well-established programs for special areas such as the Great Lakes and the Chesapeake Bay. Recently, coordinated environmental protection efforts for specific geographic areas are receiving heightened emphasis. For example, in August 1991, at the annual meeting of the Chesapeake Bay Executive Council,

EPA Administrator Reilly announced the new four-point Strategic Directions for the Chesapeake: a reexamination of nutrient reduction goals, a new focus on pollution prevention, a reinvigorated campaign to protect the Bay's living resources, and a special outreach to cultural and economic groups (CEC 1991).

- *Nitrogen Action Plan* (draft). Prompted by widespread findings of nitrate in groundwater, EPA has developed a draft action plan to address nitrogen pollution (U.S. EPA 1991d). The plan emphasizes pollution prevention, addresses both ground and surface water pollution, and also recognizes the role of atmospheric transport and deposition. The plan acknowledges the need for better scientific information, citing as one of its five major recommendations the need for more research. Research needs identified in the plan include improving our understanding of the processing of nitrogen in the environment and developing the ability to predict nitrogen transport and transformation for different land uses.
- *Agriculture in Concert with the Environment (ACE).* EPA and the Cooperative States Research Service (CSRS) of USDA established the ACE grant program to promote pollution prevention in the agricultural sector and consideration of cross-media impacts. Priority areas funded in fiscal year (FY) 1991 include a study of the national economic implications of sustainable agriculture, education and technical assistance, demonstration of promising sustainable farming practices, and research/demonstration of impacts of agricultural systems on terrestrial and aquatic species and habitat.
- *Habitat Initiative* (in preparation). In response to the SAB's identification of habitat alteration and destruction as a significant ecological risk, EPA is evaluating strategic approaches for reducing risks to habitat. Guiding principles will be a strong geographic- and risk-based focus, coordination of EPA programs, coordination with other agencies, and emphasis on immediate action.
- *Agriculture Pollution Prevention Strategy* (in preparation). Pursuant to the Pollution Prevention Act of 1990, EPA is developing a strategy for prevention of pollution from the agricultural sector. Likely areas of emphasis in the strategy are improving water quality, identifying ecologically sensitive areas, reducing pesticide use and switching to lower-risk pesticides, and conducting research on sustainable systems. EPA already has some efforts underway in these areas. For example, EPA is planning a "Public/Private Forum on Biologically Intensive Integrated Pest Management (IPM)" that will involve experts from the agricultural and environmental communities in devel-

oping recommendations to accelerate the implementation of environmentally sound pest management practices.

The Agricultural Pollution Prevention Strategy will be a comprehensive effort for the Agency and is likely to affect implementation of other programs, such as the Nitrogen Action Plan and the Pesticides in Ground Water Strategy. A major challenge in developing the Agricultural Pollution Prevention Strategy is to determine how pollution prevention concepts that have been developed primarily for the industrial sector apply to agriculture. In the industrial sector, there is a recognized hierarchy for reducing emissions of chemical pollutants to the environment (Figure 1a). The hierarchy begins with preventing the generation of pollutants via internal process modification, use of alternative inputs, etc. Next, some fraction of wastes and contaminants can be recycled or reused. Wastes that cannot be recycled are treated or controlled by traditional waste disposal methods such as incineration or discharge to water bodies. Finally, exposure to pollutants is minimized to the extent possible, for example by treatment of drinking water. In its traditional pollution control programs, EPA has been working on treat-

FIGURE 1a. Relationship of sustainable agriculture to traditional pollution prevention strategies. a. Pollution prevention hierarchy for traditional chemical pollution.

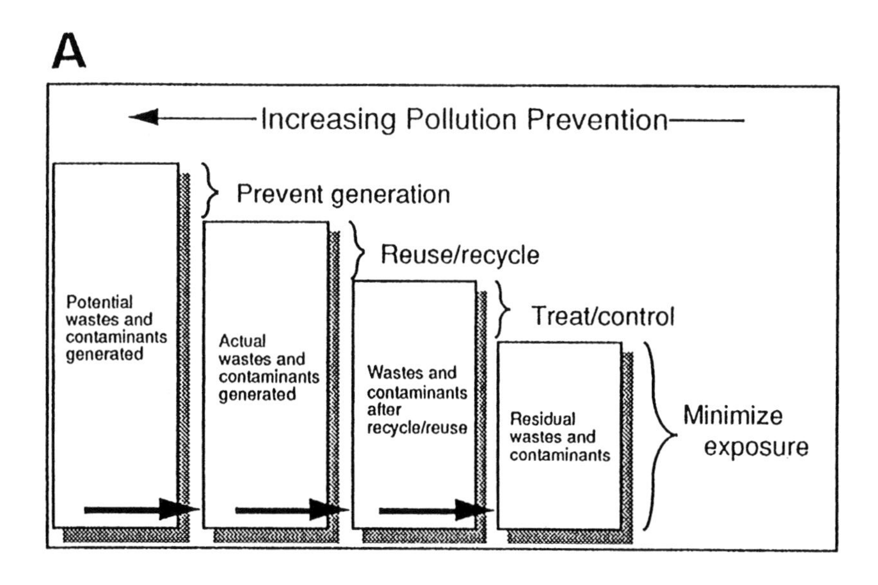

ment, control, and minimization of exposure to pollutants. The newer programs are moving toward an emphasis on pollution prevention and recycling.

How do these concepts apply to agriculture? The comparison is not straightforward, because agriculture does not simply involve discharges of contaminants to the environment. Impacts of agriculture are much broader–affecting terrestrial habitat, water quantity, soil quality, and so forth. For each of these different impacts, we must try to define what constitutes a "pollution prevention" approach (Figure 1b). Our conclusion is that the most sustainable agricultural practices are analogous to pollution prevention strategies for wastes; less sustainable agricultural practices more closely resemble waste treatment and control strategies. Sustainable agricultural practices might include, for example, reducing fertilizer runoff by adopting conservation tillage practices, avoiding use of the most sensitive habitats,

FIGURE 1b. Relationship of sustainable agriculture to traditional pollution prevention strategies. b. Conceptual approach for sustainable agriculture.

B

Increasing Sustainability

Lower impacts .... Impacts .... Higher impacts

SUSTAINABLE AGRICULTURE

Water Quality
Water Quantity
Soil Quality
Terrestrial Habitat
Biodiversity
Air Quality
Human Health
Global Climate

CONVENTIONAL AGRICULTURE

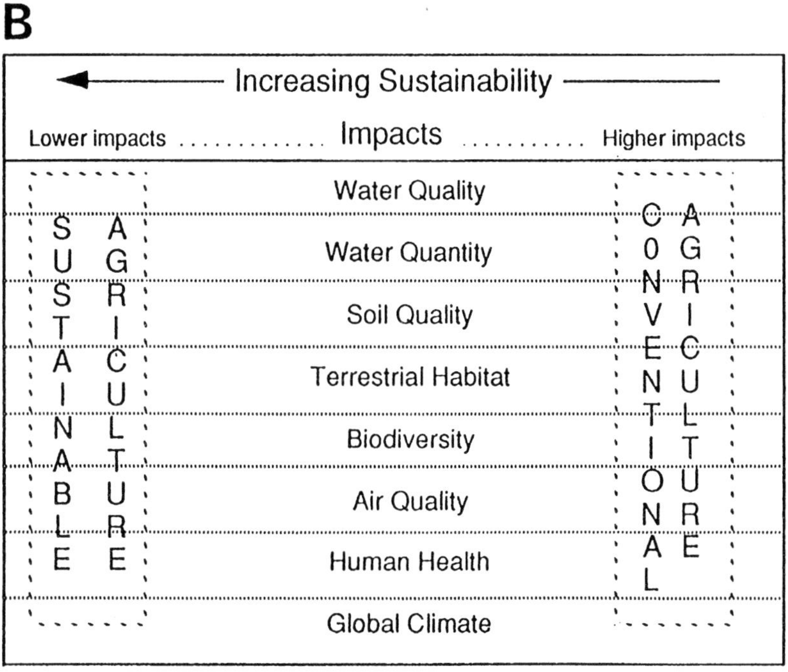

and reducing pesticide use through IPM. A less sustainable approach would rely more heavily on practices such as establishing filter strips to "control" the runoff of agricultural chemicals, creating wetlands to replace those destroyed by farming, and relying on protective clothing to prevent worker exposure to pesticides. Careful evaluation of the risks, costs, and benefits of different practices is needed to define where each would lie on the sustainability scale for each type of impact. However, there *are* many unanswered questions: What are the most sustainable agricultural practices? What are the environmental benefits? What is EPA's role in answering these types of questions?

EPA RESEARCH STRATEGY AND PROGRAMS

Agriculture-Related Research at EPA

Currently, EPA does not have a centrally planned and managed agricultural research program. Rather, agriculture-related research is undertaken by EPA's Office of Research and Development (ORD) laboratories as part of research efforts that support EPA's Office of Water, Office of Pesticides and Toxic Substances, and the Office of Air and Radiation. As indicated in Figure 2, areas of research are broad and diverse. General categories of EPA research related to agriculture are described below.

- *Terrestrial and aquatic ecosystems.* Five ORD laboratories conduct research related to the effects of agricultural chemicals and practices on terrestrial and aquatic ecosystems. One emphasis is behavior and effects of agricultural chemicals, particularly pesticides, in aquatic systems. EPA also assesses effects of wildlife exposures to agricultural chemicals in terrestrial ecosystems.
- *Agricultural chemicals in groundwater.* ORD conducts research on agricultural contamination of groundwater. One focus of research is possible remediation methods for agricultural chemicals in groundwater, including biodegradation of pesticides in aquifers and nitrate removal from groundwater through enhanced denitrification. ORD's wellhead protection research program is developing methods for delineating boundaries of and monitoring strategies for wellhead protection areas. Field testing is underway for a groundwater model that predicts leaching of pesticides to groundwater. In an effort to integrate several assessment tools, another project is developing information systems for State use in preventing groundwater contamination.

- *Biotechnology.* Objectives for ORD's biotechnology risk assessment research program (U.S. EPA 1990c) include developing methods for the detection of novel microorganisms in environmental samples, determining factors relating to growth and survival of microorganisms in the environment, developing predictive models for the transport of microorganisms in the environment, and detecting adverse environmental responses. Research topics specifically related to agriculture include reducing the amount of pesticides needed through bioregulation of the pesticide degradation process, and investigating the potential human health effects of microbial pesticides.
- *Global climate.* ORD's global climate research program includes components on understanding and predicting climate processes, biogeochemical dynamics, and ecological systems (CEES 1990). Ongoing efforts in agriculture include research on the impact of land use management practices on greenhouse gas emissions. Research projects on possible mitigation strategies are investigating enhancing agricultural soil storage of carbon, replacing fossil fuels with methane from animal waste lagoons, and using biomass as a replacement for fossil fuel. EPA is also studying the potential risks to agriculture of global warming.
- *Air pollution.* Work in this area includes effects of ozone on crops and forests, effects of pollution from alternative fuels on selected crops, and quantifying releases of volatile organic compounds, including pesticide inert ingredients.
- *Human health effects of pesticides.* Pesticide manufacturers are required to submit data from toxicological studies on the potential health effects of their products; ORD is involved in developing test methods and models used in these studies. ORD is also studying the health effects of exposure to specific pesticides and mixtures of pesticides, and conducting epidemiological studies of pesticide applicators.
- *Human exposure and risk assessment methods.* ORD develops methods to assess the potential risks of exposure to pesticides. Current areas of emphasis include food consumption patterns, exposures to pesticides in groundwater, acute exposures to food toxicants, and field worker exposures to pesticides.
- *Chemical monitoring and analysis.* Pesticide registrants are required to develop analytical methods for their products. ORD research is underway to develop less expensive methods that are easier to use, including development of immunoassays for a number of pesticides. ORD also maintains a repository of pesticide reference standards.
- *Agricultural chemical treatment and disposal.* ORD conducts re-

search on disposal methods for pesticides, maintains a database on pesticide characteristics and treatment technologies for pesticide contamination, and recently sponsored an international workshop on pesticide waste minimization, treatment, and disposal.

ORD Strategic Directions

As part of EPA's agency-wide strategic planning process, ORD developed a strategic plan that will guide research and development efforts over the next 4 years (U.S. EPA 1990b). In the past, EPA research has focused on providing support for the EPA offices developing environmental regulations. Now, ORD wants to broaden its research role and help establish EPA as a leading scientific institution capable of spearheading national and international efforts to address environmental problems. This new role is described in the key themes of ORD's strategic plan:

- *High-risk environmental problems.* ORD will enhance its technical and scientific support for Agency programs and regions, and focus research in areas of highest risk.
- *Increased emphasis on ecology and ecological assessment.* ORD will expand its ecological research efforts, including a new focus on effects at regional, continental, and global scales as well as considerations such as multiple stresses on ecosystems.
- *Innovative approaches to risk reduction and pollution prevention.* Research will focus on developing innovative technologies that can produce significant advances in pollution prevention and control.
- *Methods for determining relative risks.* Consistent and high-quality risk assessments are needed for setting national environmental priorities. ORD will develop a new generation of risk assessment methods, continue environmental health research efforts, and work toward a better understanding of human exposure.
- *Collaboration with other Federal agencies and the academic community.* To better utilize outside expertise and to make the most effective use of resources, ORD will expand cooperative research efforts with outside groups.
- *Scientific excellence and leadership.* ORD will focus its research efforts in the areas described above with the goal of becoming a nationally and internationally recognized leader in the environmental sciences.

ORD began to implement its strategic plan early in 1991, as research objectives for FY 1993 were identified. "Agroecosystems" was selected as

FIGURE 2. Summary of EPA research funding related to agriculture (about $10 million total in Fiscal Year 1991).

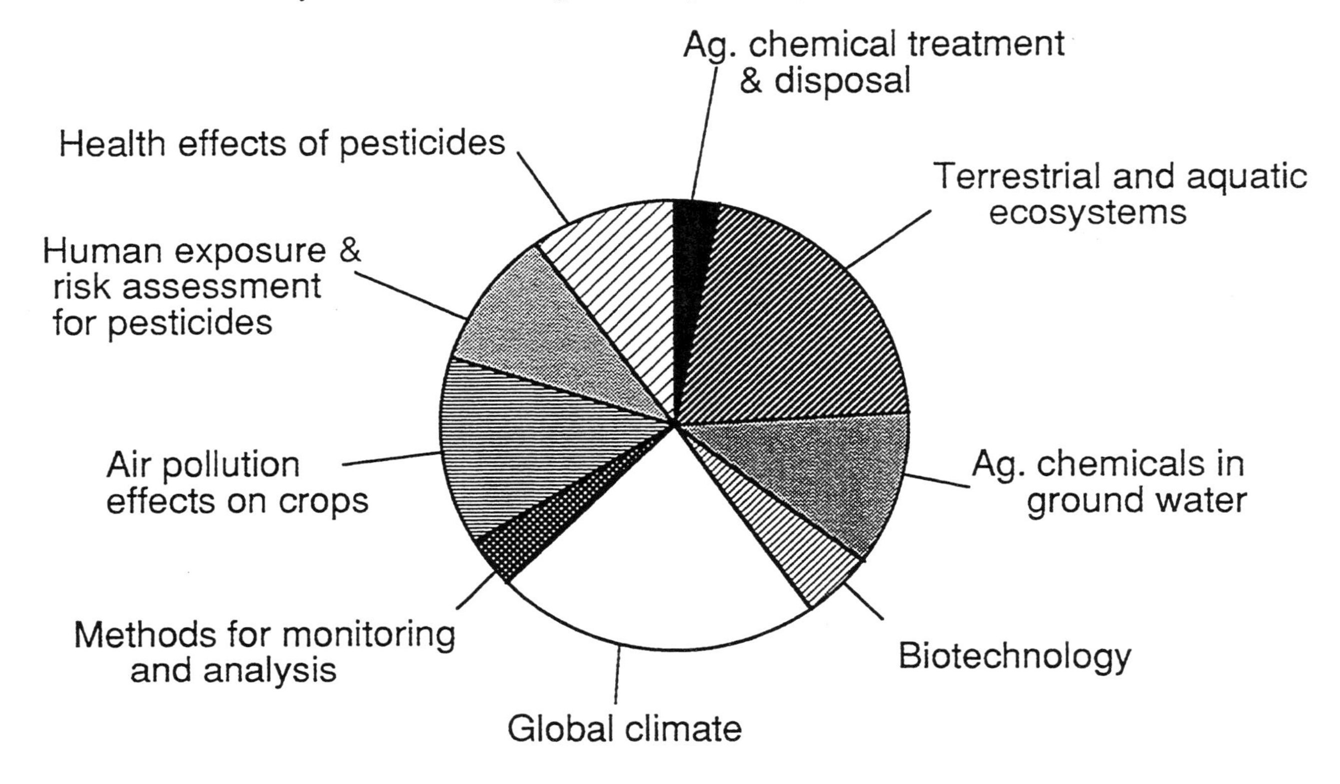

one of the topics that ORD will expand and develop into an integrated research program.

Future ORD Research

Over the next several years, ORD research related to agriculture is likely to expand significantly. An important new effort is the agroecosystems component of the Environmental Monitoring and Assessment Program (EMAP). EMAP-Agroecosystems will "develop and implement a program to monitor and evaluate the long-term status and trends of the nation's agricultural resources from an ecological perspective through an integrated, interagency process" (Heck 1991). Currently, pilot efforts are ongoing to develop a set of indicators for evaluating the health of agroecosystems (Neher, this volume).

Planning is well underway for ORD to participate, beginning in FY 1992, in the President's Water Quality Initiative (Onstad et al. 1991). In addition, ORD will work with the USDA and U.S. Geological Survey in the Midcontinent Initiative; one of the purposes of this initiative is to evaluate the effects of alternative farming practices/systems on water quality at several USDA research sites in the Midwest (Onstad et al. 1991). The EPA component of the research program (entitled Midwest Agrichemical Surface/Subsurface Transport and Effects Research, or MASTER) will focus on determining the environmental benefits of agricultural management systems, protecting and restoring ecosystem functions at the watershed level, and developing tools for implementing effective State water quality programs.

As part of the FY 1993 planning process, EPA is developing a proposal for a joint research program with USDA on integrated farm management systems. New USDA research in this area is authorized in the 1990 Farm Bill. Through this research, EPA hopes to achieve demonstrable reduction in ecological impacts within agroecosystems, to reduce chemical use and exposure through preventive measures and advanced treatment/disposal technology, and to develop regional assessments of the environmental benefits of alternative farm management systems.

RECOMMENDATIONS FOR SUSTAINABLE AGRICULTURE RESEARCH AT EPA

Research in sustainable agriculture will be useful to EPA offices as they work to implement specific programs. For example, the Office of Pesticides Programs may be able to cancel or limit use of the most harmful pesticides if sustainable agriculture alternatives to those pesticides are identified. Development of sustainable agriculture practices that can effectively reduce impacts on groundwater and surface water would also be useful to the Office of Water, which develops guidelines for State programs to protect groundwater and surface water from nonpoint source pollution.

However, focusing on specific problem pesticides or agricultural practices causing nonpoint source pollution is not the best way to develop a sustainable agriculture research program. Rather, EPA should take a systems approach to determine how agricultural practices can be modified to address environmental concerns while maintaining agricultural production capabilities. Dr. John Ikerd, in a paper presented at a recent meeting of the American Association for the Advancement of Science in New Orleans (Ikerd 1990), offered this perspective on the issue of sustainability:

> In the long run, there is no conflict between the ecologic and economic goals of sustainability. In the long run, farming systems that are not productive and profitable cannot be sustained economically no matter how ecologically sound they may be. Likewise, systems that are not ecologically sound cannot be sustained physically no matter how productive or profitable they may appear in the short run.

We suggest that EPA undertake a more integrated approach to research in sustainable agriculture; this program should include:

- *Monitoring and evaluation of environmental responses to sustainable agriculture practices.* EPA needs to develop methods to assess the environmental changes that result from the implementation of more sustainable agricultural production systems. This effort should include defining a set of environmental indicators so that the effects of alternative practices can be evaluated on a consistent basis. Potential environmental benefits to be assessed include, but are not limited to, reduction of ground and surface water contamination, improved nutrient cycling, increased soil carbon retention, improved wildlife habitat, and increased ecological diversity. EPA's EMAP is now developing a set of indicators for agroecosystem health, and data will be collected to assess national status and trends. We also need guidelines and a set of indicators that can be widely applied to other agricultural research and demonstration projects so as to provide a feedback loop to correct or adjust ineffective practices.
- *Large-scale evaluations of sustainable agriculture systems.* While USDA tends to focus on evaluating agricultural production systems at the field or farm level, EPA is more concerned with evaluating environmental effects at the watershed and regional scale. Policies to en-

courage sustainable agriculture are more likely to be implemented if we can predict the ecological benefits of widespread adoption of alternative agricultural practices. Methods are needed to evaluate the combined impacts on terrestrial and aquatic ecosystems, water and air quality, global climate, and longterm sustainability of agricultural production.

- *Development of a focus for EPA research.* Since the range of agricultural impacts on the environment is so broad and varies tremendously across the nation, and since EPA resources will be small in comparison to USDA research efforts, EPA needs to define a focus for its research so that a significant contribution can be made. Possible ways to focus EPA research include: (1) targeting specific geographic areas, (2) investigating a particular ecosystem impact, e.g., wetlands loss, (3) evaluating the environmental benefits of new, innovative agricultural production methods, and (4) establishing environmental quality standards for soils, biodiversity, etc., against which progress can be measured.
- *Working jointly with USDA.* EPA will be most effective in agricultural research if it becomes a partner with USDA in jointly planned and managed research projects. Clearly, the roles of each agency are complementary: USDA develops and demonstrates agricultural management practices, while EPA can assess the environmental responses to these practices. In the 1990 Farm Bill, USDA was charged with significant new responsibilities for research on the environmental quality impacts of agriculture. EPA should seek opportunities to work with USDA as new research programs are designed and implemented.

REFERENCES

Chesapeake Executive Council. 1991. *Chesapeake Bay Program–an action agenda.* Washington, D.C.

Committee on Earth and Environmental Sciences (CEES). 1990. *Our changing planet: The FY 1991 research plan.* Report of the Global Climate Change Research Program. Washington, D.C.: Office of Science and Technology Policy, Federal Coordinating Council on Science, Engineering and Technology.

Heck, W. W. 1991. *EMAP: Agroecosystem component.* Briefing paper for U.S. Department of Agriculture. (unpublished)

Ikerd, J. 1990. *The economics of sustainable farming systems.* Paper presented at American Association for the Advancement of Science Annual Meeting, February 15-20, 1990, New Orleans, LA.

Onstad, C. A., M. R. Burkhart, and G. D. Bubenzer. 1991. Agricultural research to improve water quality. *J. Soil Water Conserv.* 46(3):184-88.

U.S. Environmental Protection Agency. 1989. *National water quality inventory.* 1988 Report to Congress. Washington, D.C.: Office of Water Regulations and Standards.

U.S. Environmental Protection Agency. 1990a. *Reducing risk: Setting priorities and strategies for environmental protection.* Washington, D.C.: Science Advisory Board, Relative Risk Reduction Strategies Committee. SAB-EC-90-021.

U.S. Environmental Protection Agency. 1990b. *Four-year strategic research plan.* Draft. Washington, D.C.: Office of Research and Development.

U.S. Environmental Protection Agency. 1990c. *Biotechnology risk assessment research.* Washington, D.C.: Office of Research and Development.

U.S. Environmental Protection Agency. 1991a. *Strategic direction for the U.S. EPA: EPA . . . Preserving our future today.* Washington, D.C.: Office of the Administrator, U.S. EPA.

U.S. Environmental Protection Agency. 1991b. *Pesticides and ground water strategy.* Draft. Washington, D.C.: Office of Pesticide Programs.

U.S. Environmental Protection Agency. 1991c. *Proposed guidance specifying management measures for sources of nonpoint pollution in coastal waters.* Washington, D.C.: Office of Water.

U.S. Environmental Protection Agency. 1991d. *Nitrogen action plan.* Draft. Washington, D.C.: Nitrogen Work Group. 40 0/3-90/003.

Sustainable Agriculture Research and Education Program: With Special Reference to the Science of Ecology

G. W. Bird

SUMMARY. The Sustainable Agriculture Research and Education Program (formerly known as LISA) is described in relation to the Food, Agriculture, Conservation and Trade Act of 1990. A conceptual model is presented outlining the goals of sustainable agriculture and their interactions with the agroecosystem, monitored environment, enterprise controllers, and controlled system inputs. The significance of increased interaction between sustainable agriculture research and education and the science of ecology is discussed.

INTRODUCTION

United States agriculture has changed significantly during the past 50 years. These changes include increased farm size, fewer production units, increased use of synthetic fertilizers, and increased use of synthetic pesticides. During this period of time, it has also become recognized that the natural resources on which agriculture is based are finite and subject to degradation in quality. In response to the challenges associated with the changes in agriculture, Congress has legislated a research and education program in sustainable agriculture to be administrated by the U.S. Department of Agriculture (USDA).

George W. Bird is Director, Sustainable Agriculture Research and Education Program, U.S. Department of Agriculture, CSRS/SPPS, Aerospace Building, 901 D Street, SW, Washington, DC 20251.

The Sustainable Agriculture Program was initiated in 1988 under the National Agricultural Research, Extension and Teaching Policy Act Amendments of 1985. It has become widely known as LISA (Low Input Sustainable Agriculture). LISA was designed to develop agricultural procedures that are socioeconomically and environmentally compatible, and to catalyze new partnerships among farmers, nonprofit organizations, government, and academia. The program was reauthorized under Title XVI (Subtitle B) of the Food, Agriculture, Conservation and Trade Act of 1990 (FACTA) as the Sustainable Agriculture Research and Education Program (SARE). The objective of this paper is to describe the current state of SARE and its relationship to the science of ecology. This will be done through use of a conceptual model of sustainable agriculture based on the principles of systems science.

Section 1603 of the Congressional Record H11129 (October 22, 1990) defines sustainable agriculture as:

"...an integrated system of plant and animal production practices having a site specific application that will, over the long term:

- satisfy human food and fiber needs,
- enhance environmental quality and the natural resources base upon which the agricultural economy depends,
- make the most efficient use of nonrenewable resources and on-farm resources and integrate, where appropriate, natural biological cycles and controls,
- sustain the economic viability of farm operations, and
- enhance the quality of life for farmers and society as a whole" (U.S. Congress 1990).

The goals of sustainable agriculture constitute the response subsystem of a conceptual model of sustainable agriculture (Figure 1). The other two subsystems of this model are the external environment and object of control. Within the external environment subsystem, the monitored environment includes factors such as light, temperature, water, soil texture, biologicals, chemicals, economics, and social mandates. The farm family, board of directors, and farm manager are categorized as enterprise controllers. The controlled inputs include water, chemicals, biologicals, labor, technology, and energy. The object of control subsystem is the farm, consisting of various atmospheric resources, organic resources (plants, animals, and microbes), sedimentary resources, and the individual responsible for the current state of the production system (farmer). Within the object of control subsystem there is continual interaction of organisms and groups of organ-

FIGURE 1. Conceptual model of sustainable agriculture as defined in Section 1603 of the 1990 Farm Bill [Congressional Record H11129 (22 Oct 90)].

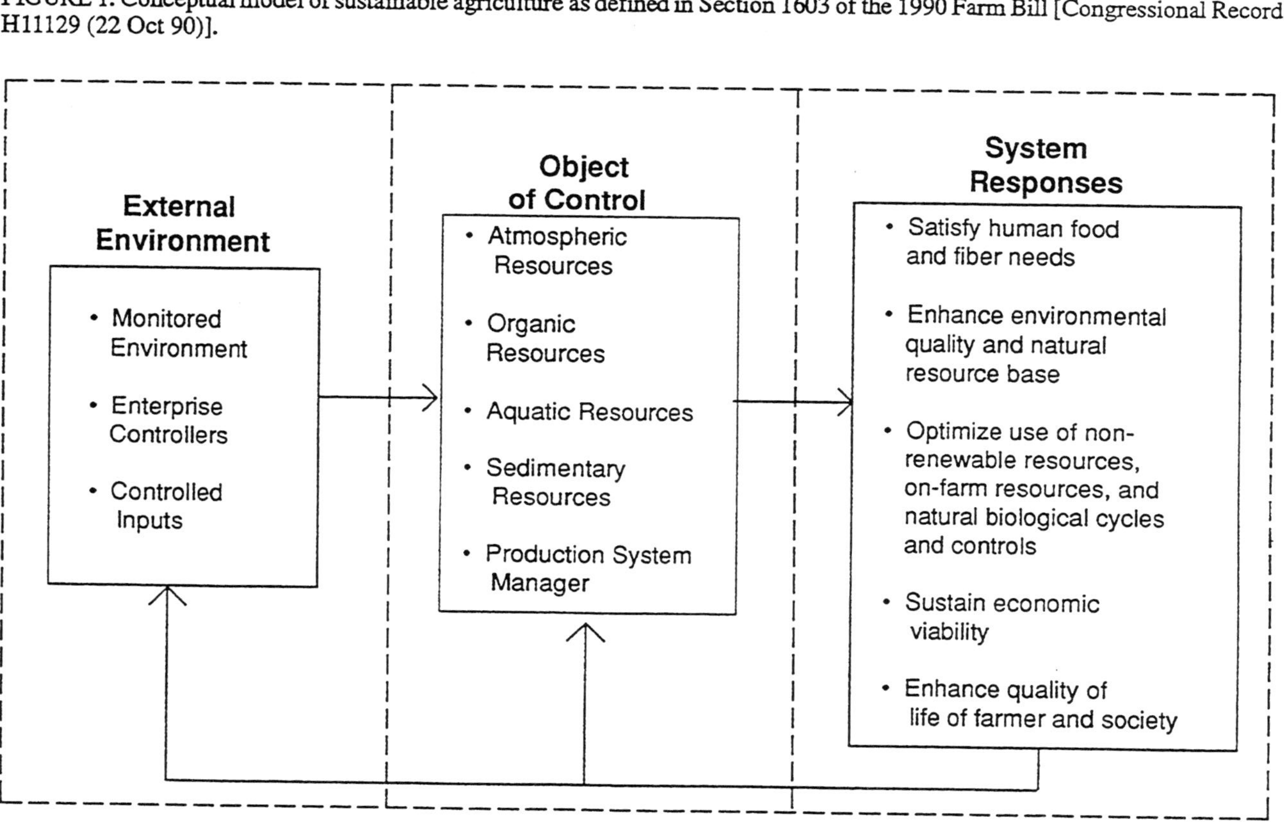

isms with their physical environment. This provides a direct interface with the science of ecology.

SUSTAINABLE AGRICULTURE RESEARCH AND EDUCATION PROGRAM

The purpose of SARE is to encourage research designed to increase our knowledge concerning agricultural production systems that:

- maintain and enhance the quality and productivity of the soil;
- conserve soil, water, energy, natural resources, and fish and wildlife habitat;
- maintain and enhance the quality of surface and ground water;
- protect the health and safety of persons involved in the food and farm system; and
- increase employment opportunities in agriculture (FACTA, Title XVI, Subtitle B, Sec. 1619).

Under SARE, USDA is responsible for facilitating and increasing scientific investigation and education in order to:

- reduce, to the extent feasible and practicable, the use of chemical pesticides, fertilizers, and toxic materials in agricultural production;
- improve low-input farm management to enhance agricultural productivity, profitability, and competitiveness;
- promote crop, livestock, and enterprise diversification (FACTA, Title XVI, Subtitle B, Sec. 1619).

In addition, SARE charges USDA with facilitating projects to:

- study, to the extent practicable, agricultural production systems that are located in areas that possess various soil, climate, and physical characteristics;
- study farms that have been, and will continue to be, managed using farm production practices that rely on low-input and conservation practices;
- take advantage of the experience and expertise of farmers and ranchers through their direct participation and leadership in projects;
- transfer practical, reliable, and timely information to farmers and ranchers concerning low-input sustainable farming practices and systems; and

- promote a partnership between farms, nonprofit organizations, agribusiness, and public and private research and extension institutions (FACTA, Title XVI, Subtitle B, Sec. 1619).

SARE is administrated as a competitive research and education grants program. More than 850 proposals were evaluated during the first 3 years of operation. A total of 154 projects were funded, representing approximately 18 percent of those submitted. Congress appropriated a total of $12.8 million for the first 3 years, and the funded projects provided matching contributions of $14 million. In fiscal year 1991, SARE was funded at $6,725,000. An additional $1 million was made available for a new grants program (Agriculture in Concert with the Environment) which is administrated jointly with the U.S. Environmental Protection Agency.

The research and education program is administrated through four Regional Administrative Councils. This procedure was codified in FACTA, and a National Sustainable Agricultural Advisory Council (NSAAC) was mandated. The responsibilities of NSAAC are to:

- make recommendations to the Secretary concerning research and extension projects that should receive funding;
- promote the programs established under this chapter at the national level;
- coordinate research and extension activities funded under such programs;
- establish general procedures for awarding and administering funds under this chapter;
- consider recommendations for improving such programs;
- facilitate cooperation and integration between sustainable agriculture, national water quality, integrated pest management, food safety, and other related programs; and
- prepare and submit an annual report (FACTA, Title XVI, Subtitle B, Sec. 1621).

FACTA stipulates that NSAAC include representatives of the following organizations or categories:

- the Agricultural Research Service
- the Cooperative State Research Service
- the Soil Conservation Service
- the Extension Service

- State cooperative extension services
- State agricultural experiment stations
- the Economic Research Service
- the National Agricultural Library
- the Environmental Protection Agency
- the Farmers Home Administration
- the Board on Agriculture of the National Academy of Sciences
- private nonprofit organizations with demonstratable expertise
- farmers utilizing systems and practices of sustainable agriculture
- the United States Geological Survey
- agribusiness
- other specialists in agricultural research or technology transfer, including individuals from colleges eligible to receive funds under the Act of August 20, 1890, or other colleges or universities with demonstrable expertise (FACTA, Title XVI, Subtitle B, Sec. 1621).

The responsibilities of the existing Regional Administration Councils are to:

- make recommendations to the Advisory Council concerning research and extension projects that merit funding;
- promote the programs established at the regional level;
- establish goals and criteria for the selection of projects authorized within the applicable region;
- appoint a technical committee to evaluate the proposals for projects to be considered by such council;
- review and act on the recommendations of the technical committee; and
- prepare and make available an annual report concerning funded projects with an evaluation of the project activity (FACTA, Title XVI, Subtitle B, Sec. 1621).

Members of the Regional Administrative Councils include representatives from:

- Agricultural Research Service
- Cooperative State Research Service
- Extension Service
- State cooperative extension services
- State agricultural experiment stations
- Soil Conservation Service

- State departments engaged in sustainable agriculture programs
- nonprofit organizations with demonstrable expertise
- farmers utilizing systems and practices of sustainable agriculture
- agribusiness
- State or United States Geological Survey
- other persons knowledgeable about sustainable agriculture and its impact on the environment and rural communities (FACTA, Title XVI, Subtitle B, Sec. 1621).

The Regional Administrative Councils have developed protocols for establishing regional priorities, inviting proposals, evaluating proposals, and recommending projects for funding. The four Regional Councils use Technical Review Committees for comprehensive evaluations of all proposals. In general, the following six types of projects are eligible for funding:

- Education
- On-Farm Demonstration
- Component Research
- Exploratory Research
- Integrated Systems Research
- System Impact Assessment.

Priority has been given to projects in the integrated systems category, with special reference to either whole-farm or ecosystem analysis. Impact assessment projects include economic, social, and environmentally oriented research and education activities. Through April of 1991, 33 whole-farm/ ecosystem projects, 27 education-demonstration, 44 component research, and 8 impact assessment projects have been funded (Table 1). The results of the projects are becoming available, and examples are featured in the 1990 SARE Annual Report (Madden et al. 1991).

CURRENT STATE OF SUSTAINABLE AGRICULTURE

Interest in sustainable agriculture has increased greatly during the past 5 years. Agricultural journals have discussed many aspects of LISA, and the term has become widely recognized by the agricultural community. A significant number of nonprofit organizations related to sustainable agriculture have emerged. Many of these are farmer-initiated and farmer-managed. Some have merged to form multiorganizational coalitions.

New emphasis has been given to sustainable agriculture through the

TABLE 1. Types of USDA Sustainable Agriculture Research and Education projects funded (1988-April 1991) (Madden et al. 1991).

	Project Categories			
Commodities-Subcategories	Whole Farm-Ecosystem Studies	Education-Demonstration Projects	Component Research Projects	Impact Assessment Projects
Crops:				
Fruits and Nuts	7	-	5	-
Vegetables	7	2	1	-
Field Crops/Rotations	5	-	5	-
Cover Crops	1	1	2	-
Weeds and Tillage	-	-	7	-
Disease Control	-	-	3	-
Livestock:				
Crop-Livestock	11	-	1	-
Livestock	2	-	-	-
Pasture Management	-	1	2	-
Nutrient and Waste Management	-	-	6	-
Manure/Compost Management	-	-	7	-
Impact Studies:				
Environmental Impact	-	-	-	2
Farm Budgets	-	-	-	4
Market-level Economic and Social	-	-	-	2
Other Educational and Demonstration:				
Information Systems/Databases	-	5	-	-
Educational Materials	-	8	-	-
Conferences	-	5	-	-
Demonstration Projects	-	5	-	-
Miscellaneous	-	-	5	-
Totals	33	27	44	8

creation of a broad array of public sector centers, academic positions, and special programs. There has also been a distinct increase in the publication of reviews, books, scientific articles, and journals dedicated to sustainable agriculture.

Role of the Science of Ecology in Sustainable Agriculture

The science of ecology has much to offer sustainable agriculture. Although ecology is a relatively young discipline, its principles, language, and methodologies have become relatively well developed during the past 45 years. They appear to be exceptionally useful as a framework for the research and education needs of sustainable agriculture. The language of ecology mandates a thought process that is extremely compatible with the objectives of sustainable agriculture. Unfortunately, this is not always true of the disciplinary, component, reductionist, and system-input research and education activities related to the industrial model of commercial agriculture that has emerged during the past 45 years.

Fundamental concepts of ecosystems such as food chains, food webs, carrying capacity, behavior of energy, and ecological efficiencies are directly related to all agricultural production systems. For example, most daily farming activities are impacted by a combination of Liebig's Law of the Minimum and Shelford's Law of Tolerance (Smith 1990). Although the subdiscipline of agroecology is growing, agriculture has not yet widely adopted its language and procedures. Formal integration of many aspects of the science of ecology into agricultural science would go a long way towards increasing the probability of achieving the goals of sustainable agriculture in a timely manner.

CONCLUSION

Agriculture has undergone continual change since its evolution 10,000 years ago. It will continue to be highly dynamic. Although many aspects of agriculture are regional or even site specific, the principles of the science of ecology pertain to all agroecosystems. They must be used, however, in a context that includes research, technology development, and appropriate diffusion of innovations (Rogers and Shoemaker 1971; Zaltman and Duncan 1977) designed for the timely adoption and maintenance of systems of sustainable agriculture (Figure 2).

During the past 40 years, it has been evident that the natural resources used in agricultural production systems are both finite and subject to decline in quality. Social institutions, however, resist change unless strong incentives are present. Sustainable agriculture is likely to have an important impact on interactions among our institutions, production systems and natural resources (Figure 3).

Of even greater significance, sustainable agriculture has an opportunity to play a major pioneering role in the evolution of a future era of sustainable development. Adoption of the concepts, methodologies, and language of the science of ecology by sustainable agriculture should contribute significantly

FIGURE 2. Conceptual model of science, technology, and diffusion of innovations related to the adoption of sustainable agriculture.

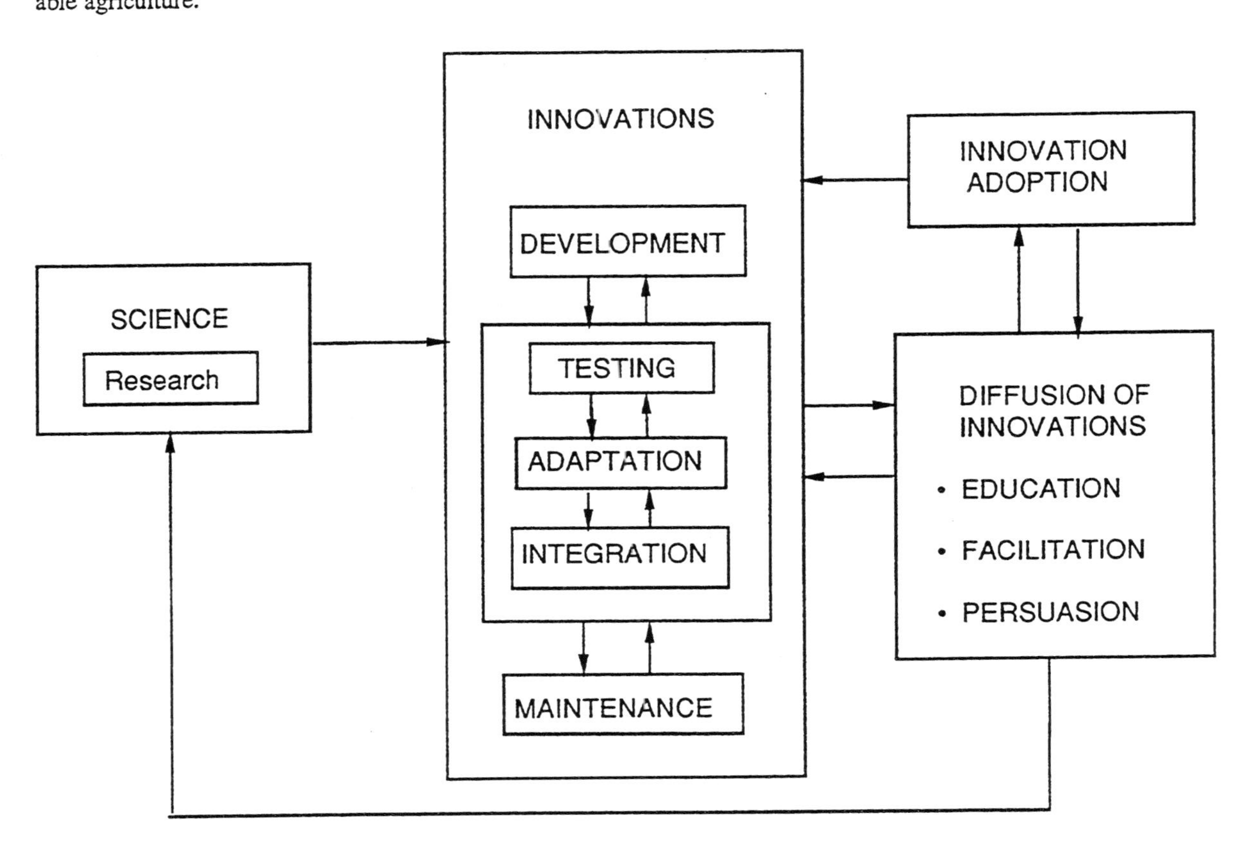

FIGURE 3. Conceptual model of the interactions among social institutions, production systems, and natural resources.

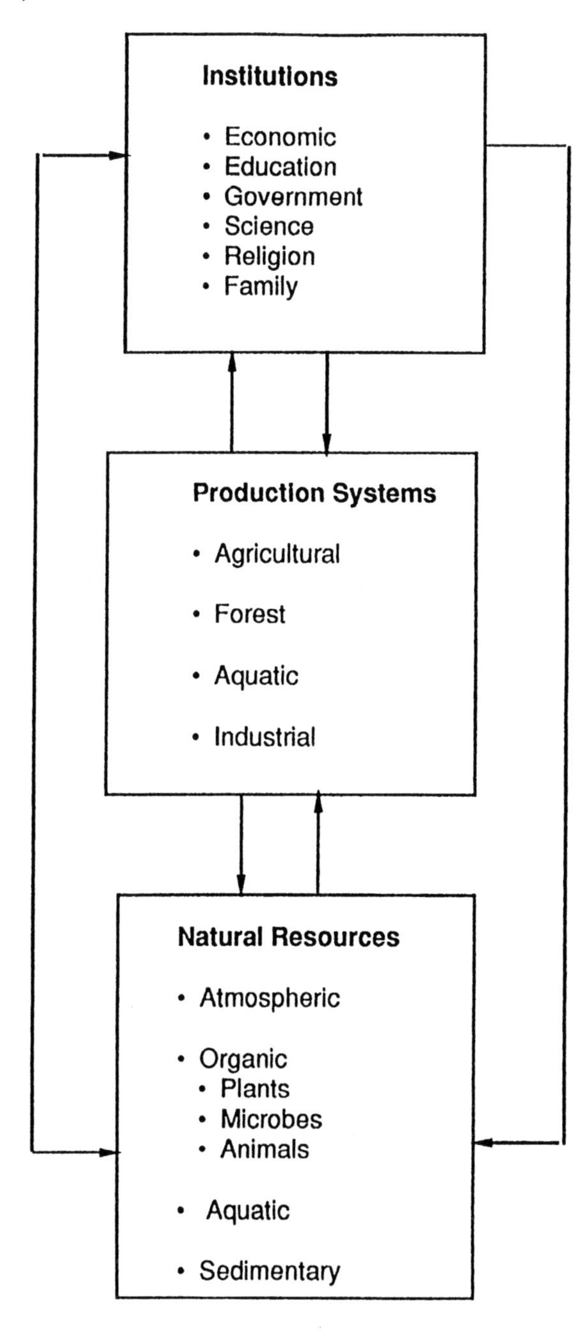

to this goal. It appears that SARE may be an important catalyst in this process.

REFERENCES

Madden, J. P., W. H. Brown, F. R. Magdoff, D. E. Schlegel, and S. S. Waller. 1991. *1991 Annual progress report: Sustainable Agriculture Research and Education Program*. USDA/CSRS/SARE. Washington, D.C. 21 pp.

Rogers, E. M., and F. F. Shoemaker. 1971. *Communication of innovations*. New York: The Free Press. 476 pp.

Smith, R. L. 1990. *Ecology and field biology*. New York: Harper and Row Publishers. 922 pp.

U.S. Congress. House. 1990 Farm Bill. Congressional Record. (22 October 1991). H11129.

Zaltman, G., and R. Duncan. 1977. *Strategies for planned change*. New York: John Wiley & Sons. 404 pp.

Precollege Education: A Vital Component If Sustainable Agriculture Is to Take Root

Sandra Henderson

SUMMARY. Individuals are growing increasingly concerned over the state of the environment and are aware that many of their current choices and actions will have lasting impacts on environmental health. However, the general lack of science/environmental education in the United States has resulted in a public that is poorly prepared to understand these problems and formulate effective solutions. Proponents of sustainable agriculture must recognize that the precollege school system (K-12) is an essential place to begin disseminating information that will allow individuals to understand the implications of different agricultural practices in terms of their own health, the health of their environment, and the maintenance of their food supply. Science education in the United States is undergoing reform, providing an excellent opportunity to use sustainable agriculture as a theme in teaching science. Efforts are underway to make science more relevant

---

Sandra Henderson is Senior Scientist with ManTech Environmental Technology, Inc., USEPA Environmental Research Laboratory, 200 SW 35th Street, Corvallis, OR 97333.

The research described in this article has been funded by the U. S. Environmental Protection Agency. This document has been prepared at the EPA Environmental Research Laboratory in Corvallis, OR, through contract #68-C8-0006 to ManTech Environmental Technologies, Inc. It has been subjected to the Agency's peer and administrative review and approved for publication. Mention of trade names or commercial products does not constitute endorsement or recommendation for use.

The author gratefully acknowledges Helene Murray, Project Leader of the Min-to-Brown Agricultural Project, and Margaret Krome of the Wisconsin Rural Development Center for their input on the case study section of this paper. In addition, appreciation is due to Jim Barrett, Steve Holman, Tom Moser, Helene Murray, and Richard Olson for their insightful reviews and comments.

relevant to real world situations through the establishment of partnerships between subject matter experts and educators. Scientists, farmers, and teachers can form alliances that will enhance students' knowledge of both sustainable agriculture and science.

INTRODUCTION

Proponents of sustainable agricultural practices recognize that education must play a prominent role if such practices are to be successfully introduced and implemented in the United States (Keeney 1989; NAS 1989; Francis et al. 1990). However, current educational efforts are generally limited to information transfer among researchers, farmers, extension agents, and other members of the agricultural community (Francis et al. 1990; Bird, this volume). While such efforts are to be encouraged, it is critical that educational efforts be expanded to include the largest segment of society: the nonfarm public (McGrath 1990).

There are two main reasons why it is important that all citizens have a basic understanding of the concepts and issues in sustainable agriculture. First, the public influences government agricultural policies through its choice of representatives and through communications (e.g., letter writing, town meetings) with elected and nonelected officials. Much has been written about the importance of government policies in the development of U.S. agriculture (NAS 1989). Federal polices influence the types and amounts of crops grown, the availability of specific pesticides, the extent to which various soil conservation practices are adopted, and other aspects of agriculture related to sustainability. At the state and local levels, right-to-farm laws and land use decisions may greatly affect agriculture. Although the influence of some government policies on farm practices may be overestimated (Doering, this volume), agricultural policies will have an important influence on the direction of U.S. agricultural practices.

Second, individuals make food choices. As consumers, individual choices, (e.g., accepting blemished fruit and vegetables that were produced using reduced chemical input) will affect the direction of agriculture. Gussow and Clancy (1986) refer to consumers as the "final arbiter of the food system." Consumers need to make food choices that not only enhance their own health, but also contribute to the protection of the environment, and to the development of a sustainable agriculture. The effect of the Alar scare on segments of the U.S. apple industry is just one example of the cumulative effects of many individual consumer choices on agriculture (McCullagh 1989).

The precollege school system (K-12) is an essential place to begin disseminating information that will allow individuals to understand the implications of different agricultural practices in terms of their own health, the health of their environment, and the maintenance of their food supply. Precollege education provides the only formal education that the majority of the public receives. Most students who do go on to college receive little education at that level on agriculture and related topics. If an agriculturally literate populace is a goal, then precollege education must be the primary mechanism for achieving that goal.

The timing is optimal for developing programs that address sustainable agriculture for precollege students because (1) there is strong concern about the general state of education in the United States, (2) environmental issues are highly visible, (3) sustainability is emerging as a central theme in resource use and management, and (4) there is a search for societal relevance in science course materials (Hart and Robottom 1990; DeBoer 1991; NAS 1988).

This paper will address the following topics: (1) agricultural education in the United States, (2) sustainable agriculture as a topic in science education, (3) scientist/educator partnerships, and (4) examples of effective precollege sustainable agriculture programs.

AGRICULTURAL EDUCATION IN THE UNITED STATES

The National Research Council (NRC) reported that the majority of U.S. students enter school knowing little about agriculture and then graduate from high school only slightly better informed (NAS 1988). Not surprisingly, the limited number of agriculture classes are generally found in rural schools. Yet most American schoolchildren are urban rather than rural. Students in urban areas lack first-hand knowledge or experience of agriculture and rural life, and have little exposure to agricultural education in the school curriculum (Phipps and Osborne 1988).

Traditionally, precollege agricultural education has focused on vocational training designed to prepare individuals for careers in agriculture. This education in agriculture has a limited audience. Education about agriculture, which could have a much larger audience, is largely lacking in our public school systems with the result that we are generally a nation that knows little about the sources of our food, fuel, and fiber (NAS 1988; Phipps and Osborne 1988).

While the NRC report does not refer specifically to sustainable agriculture, general agricultural literacy is a key foundation in supporting the movement toward sustainable practices. The general populace needs basic knowledge to make informed electoral and consumer decisions regarding alternative forms of agriculture. Students who will become specialists in the many fields that constitute sustainable agriculture (e.g., agronomy, ecology, sociology, economics) need a common language in order to communicate.

SUSTAINABLE AGRICULTURE AS A THEME IN SCIENCE EDUCATION

The existing time pressures and subject requirements in most school districts make it almost impossible to add new courses (Collette and Chiapetta 1989). In light of this, integrating sustainable agriculture concepts into existing courses is a more viable alternative to pursue. Sustainable agriculture can be taught within a number of school subjects including science, social studies, economics, and political science. However, this discussion will focus on incorporating sustainable agriculture concepts into existing science education programs.

Science education in the United States is in a state of reform as efforts are underway to make science more relevant to real world situations (FCCSET 1991; AAAS 1989; DeBoer 1991). The reasons for reform include (1) outdated teaching materials, (2) poorly trained teachers, (3) lack of subject relevance, (4) lack of multidisciplinary approaches, and (5) the difficulties in teaching complex issues that are fraught with uncertainty (Holdzkom and Lutz 1989). A new movement, Science-Technology-Society (STS), is trying to find a solution to what many educators believe is a crisis in school science: a disjuncture between school science and the kind of science background needed by citizens making decisions in a technological society (Hart and Robottom 1990). STS stems from a position statement by the National Science Teachers Association (1983) advocating new goals for science education that produce scientifically literate citizens. These goals include conveying to students the knowledge, skills, technology, values, and ethics needed to live in a science- and technology-based society (Collette and Chiapetta 1989).

The concern over the state of science education is coupled with an overall sense of the need for improved environmental education. An effective response to complex environmental issues requires an understanding of the scientific foundation inherent in all ecological concerns. The National Environmental Education Act (PL 101-619), in finding that previous environmental education efforts have not been adequate, mandates that actions will be taken to encourage public- and private-sector involvement in the development of curricula, teacher training, and other activities to improve awareness of environmental problems.

Development of sustainable agriculture is based on the recognition that agricultural landscapes are, in fact, ecological systems (Carroll et al. 1990). Therefore, sustainable agriculture is a good avenue for teaching ecological principles. Specific topic areas of the ecological or biological sciences that can be taught in an agricultural context include energy flow, applied genetics, cycles (e.g., nitrogen), water balance, plant physiology, ecophysiology, species diversity, landscape ecology, predation, and competition (NAS 1988).

SCIENTIST/FARMER/EDUCATOR PARTNERSHIPS

It will not be enough for experts in sustainable agriculture to develop teaching materials or programs and "give" them to teachers for implementation. The development of partnerships between active proponents of sustainable agriculture and educators will be crucial to successful integration of sustainable agriculture curricula and activities into classrooms. Educational research clearly shows that a lack of teacher input is the primary barrier to the successful implementation of curricula (DeBoer 1991). When curricula are developed by experts (e.g., sustainable agriculture researchers and university agriculture educators) and then given to teachers, they are not nearly as likely to be used as when true partnerships among scientists, farmers, and teachers are formed to capitalize on the expertise of each. When educators develop "ownership" of the eventual product (i.e., curriculum), the program has a better guarantee of success (DeBoer 1991). Scientists and farmers will bring to the partnership their knowledge and understanding of sustainable agriculture systems and issues as well as their ideas for student activities and experiments. The teachers will contribute expertise in science teaching, student needs, and curriculum design.

Partnerships and collaborations are considered vital to the success of educational reform (FCCSET 1991; AAAS 1989). Many agencies and environmental groups have recognized the need for closer communication between research scientists, technical experts, and educators, but in practice the linkage has never been strong.

The following recommendations should be considered in the formation of partnerships for the purpose of developing precollege curricula in sustainable agriculture:

1. Agriculturalists and teachers should identify ways to develop joint programs. Successful educational programs will include representa-

tives of all interested parties at the outset of the project. Researchers should not tell educators what they need to know or how to teach. Researchers have a responsibility to articulate their results in a way that can be applied and understood by teachers.

2. Develop all teaching materials with the goal of encouraging critical thinking. Emphasize scientific inquiry and process.
3. "Hands-on" activities should be emphasized whenever possible. Research has demonstrated that the learning process is enhanced through first-hand experience.
4. Because resources are often limited, find out what material is available that may be adaptable for sustainable agriculture education. For example, programs such as "Agriculture in the Classroom" and "The Growing Classroom" could provide useful background information and ideas (NAS 1988).
5. Provide opportunities for teachers to work at sustainable agriculture research institutes and centers during the summers. This would allow the teachers to become familiar with sustainable agriculture practices and concepts.
6. When teaching materials have been developed, use innovative ways to disseminate them. For example, the teachers directly involved in the development of the teaching materials could hold demonstrations or workshops for other teachers at inservice days, allowing new teachers the opportunity to be trained in the use of the materials.
7. Establish an informational network that will allow the exchange of ideas between researchers, farmers, and educators. This network will be particularly useful once teaching materials are in use. Encourage the exchange of ideas and suggestions for improvement from the practitioners.
8. It is important to note that the goal of education is not to tell students what to think. Even though proponents of sustainable agriculture are committed to implementing sustainable practices, it is inappropriate to use the school system for advocacy. Instead, help students learn how to think critically about the issues. It follows that informed individuals will make informed decisions.

EXAMPLES/CASE STUDIES

The following are examples where, through partnerships and collaborative efforts, teaching materials addressing the principles of sustainable agriculture have been developed for precollege level education.

Minto-Brown Agricultural Project

Minto-Brown Island Park, near the Willamette River, is an 892-acre park that is owned by the City of Salem, Oregon. The island is managed for multiple uses such as farmland, recreation, and wildlife refuge. Approximately 300 acres of the park land is leased and farmed commercially. Recent concerns over pesticide use by the tenant farmers led to the creation of the Minto-Brown Park Task Force. The concerns of the task force include pesticide use in the park, but go beyond that concern to address the question of how agriculture at the urban fringe will survive as Salem expands: Can the needs of people, natural plant and wildlife communities, and agriculture be met adequately on the same piece of land? The task force is composed of individuals representing diverse interests, including farmers, researchers, state agencies, school districts, environmental groups, nonprofit agriculture groups, and the agricultural industry.

The objectives of the task force are two-fold: education and research. It was determined that the agricultural aspects of the park provided an excellent opportunity to expose local grade school children to both conventional and sustainable agricultural practices. Currently, teaching materials are being developed through partnerships with scientists, educators, and farmers. The goals of the educational effort are to (1) encourage critical thinking, (2) allow students to experience agriculture first-hand, (3) expose students to the multiple-use resource perspective and its inherent complexities, (4) demonstrate concepts and practices of both sustainable and conventional agriculture to students, and (5) provide "hands-on" activities.

Sustainable Agriculture Curriculum Project

Toward a Sustainable Agriculture: A Teacher's Guide (SACP 1991) is a unique curriculum targeted at high school agriculture instructors with the intent that it also be used by science and social studies instructors. This project was sponsored by an interagency task force in Wisconsin and is being distributed by the Wisconsin Rural Development Center. Farmers, researchers, and high school teachers formed a partnership to develop the curriculum. The goal of the project was to develop a creative, action-oriented curriculum that would interest and engage both students and teachers while emphasizing systems thinking, cooperative learning, and concept learning.

The instruction manual for the curriculum is divided into three sections: (1) a teacher reference guide that provides background information, (2) an instruction unit that can be used as a course planning tool, and (3) a collection of 21 learning activities for student use. Topics suitable for science instruction include composting, nitrate leaching, weed management, insect pest management, plant disease, and pesticide effects on groundwater. Also addressed are the economic, political, and historical aspects of sustainable agriculture through topics such as community development, career options, and market development.

CONCLUSIONS

Those interested in sustainable agriculture should not overlook the importance of providing information and educational materials to the precollege school system. With the general reform of the Nation's educational system, and science in particular, the opportunity is available for including teaching materials on a topic that will continue to grow in importance and relevance: sustainable agriculture. The environmental challenges for the next generation are enormous in their magnitude and complexity. We need to prepare the next generation for the decisions that lie ahead as they make consumer choices and fashion national policies addressing human health.

REFERENCES

American Association for the Advancement of Science. 1989. *Science for all Americans*. Washington, D.C.: AAAS. 217 pp.

Carroll, C. R., J. H. Vandermeer, and P. M. Rosset, eds. 1990. *Agroecology*. New York: McGraw-Hill. 641 pp.

Collete, A. T., and E. L. Chiappetta. 1989. *Science education in the middle and secondary schools*. Columbus, OH: Merrill Publishing Co. 471 pp.

DeBoer, G. E. 1991. *A history of ideas in science education*. New York: Teachers College Press. 268 pp.

FCCSET Committee on Education and Human Resources. 1991. *By the year 2000: First in the world*. Washington, D.C. 321 pp.

Francis, C., J. King, J. DeWitt, J. Bushnell, and L. Lucas. 1990. Participatory strategies for information exchange. *Am. J. Altern. Agric.* 5(4):153-60.

Gussow, J. D., and K. L. Clancy. 1986. Dietary guidelines for sustainability. *J. Nutr. Educ.* 18(1):1-5.

Hart, E. P., and I. M. Robottom. 1990. The science-technology movement in science education: A critique of the reform process. *J. Res. Sci. Teaching* 27(6):575-88.

Holdzkom, S., and P. B. Lutz, eds. 1989. *Research within reach: Science education*. Washington, D.C.: National Science Teachers Association. 216 pp.

Keeney, D. R. 1989. Towards a sustainable agriculture: Need for clarification of concepts and terminology. *Am. J. Altern. Agric.* 4 (3/4):101-105.
McCullagh, J. C. 1989. A mother's crusade. *Org. Garden.* 36(4):32-37.
McGrath, D. 1990. Two way communication. *Pacific Northwest Sustainable Agriculture* 2(4):7.
National Academy of Science. 1988. *Understanding agriculture: New direction for education.* Washington, D.C.: National Academy Press. 68 pp.
National Academy of Science. 1989. *Alternative agriculture.* Washington, D.C.: National Academy Press. 448 pp.
National Science Teachers Association. 1983. NSTA position statement on science-technology-society: Science education for the 1980's. In *Science teaching: A profession speaks*, ed. F. K. Brown and D. P. Butts, 294-99. Washington, D.C.: National Science Teachers Association.
Phipps, L. J., and E. W. Osborne. 1988. *Handbook on agricultural education in public schools.* Danville, IL: The Interstate Printers and Publishers. 585 pp.
Sustainable Agriculture Curriculum Project. 1991. *Toward a sustainable agriculture: A teacher's guide.* Madison, WI: University of Wisconsin. 151 pp.